Hazardous Materials Unveiled

Eric J. Neal

Published by Eric Neal, 2025.

HAZARDOUS MATERIALS UNVEILED

First edition. February 26, 2025.

ISBN: 979-8230355519

Written by Eric J. Neal.

Also by Eric J. Neal

From the Streets to a Chief
Igniting Excellence: Challenges of a First-Year Fire Chief
Echoes From the Ashes
The Roadmap to Fire Behavior
The Suppression Lesson: Water vs Fire
From the Shadows: SMU's Journey
Hazardous Materials Unveiled
Rescue Operations Strategies for Saving Lives
Firefighter Survival: Tactics for Staying Alive
Incident Command and Leadership Decision Making on the Fireground
Apparatus & Equipment Mastery Tools of the Trade
Smoke Between Us
Firefighter EMS & Medical Aid in the Fire Service

Table of Contents

To the brave men and women who risk their lives to protect others—firefighters, first responders, and hazardous materials specialists. Your dedication, sacrifice, and commitment to safety inspire us all. This book is for you.

Hazardous Materials Unveiled: The Foundations of Safety and Response

Chapter 1: Introduction to Hazardous Materials

(Awareness Level)

Understanding hazardous materials is vital for individuals involved in their management, whether students, firefighters, first responders, or professionals in related fields. The inherent risks these materials pose to life, health, property, and the environment demand a clear comprehension of their chemical and physical properties. Such understanding is crucial due to their potential to cause harm through various routes of exposure like inhalation, skin absorption, ingestion, or direct contact. Recognizing the difference between hazardous materials and hazardous waste further highlights the need for precise definitions, as each requires different safety measures and protocols. This distinction becomes especially critical when planning and implementing effective strategies to mitigate risks.

This chapter delves into several core areas concerning hazardous materials, starting with their classification under prominent systems such as the Department of Transportation's categories, the National Fire Protection Association's Fire Diamond, and the Globally Harmonized System. Each system plays a crucial role in identifying, handling, and responding to these substances effectively, ensuring uniformity across transportation, labeling, and emergency responses. The importance of specialized HazMat training for emergency responders is also underscored, emphasizing its role in equipping personnel with the necessary skills to handle incidents safely and efficiently. Additionally, the chapter explores the legal landscape governing hazardous materials, focusing on key regulations enforced by bodies like OSHA, DOT, and EPA. These laws are designed to protect public health and the environment while ensuring compliance and safety. By weaving together these elements, the chapter provides a comprehensive introduction to hazardous materials, aiming to enhance preparedness and the capability to manage potential hazards effectively.

Definition of Hazardous Materials

Understanding hazardous materials is crucial for anyone involved in their handling, whether as a student, firefighter, first responder, or professional in related fields. Hazardous materials pose significant risks to life, health, property, and the environment due to their chemical and physical properties. They are defined by their potential to cause harm through various exposure routes, such as inhalation, skin absorption, ingestion, and direct contact.

Hazardous materials are often confused with hazardous waste; however, these two terms represent different concepts. While both can be dangerous, hazardous materials are substances that are actively used or transported, posing immediate threats due to their inherent properties. On the other hand, hazardous waste refers to substances that are disposed of after being used, still retaining the potential to cause environmental or health issues if not managed properly. Recognizing this distinction is essential when planning safety measures and protocols.

There are several common types of hazardous materials that students and professionals need to identify. Toxic substances, which include poisons, are one category. These materials can cause severe health effects, ranging from acute symptoms like dizziness and nausea to chronic conditions such as organ damage and cancer, depending on exposure level and duration. For example, lead and mercury are toxic substances with long-term health implications.

Flammable liquids like gasoline represent another category. These materials have low flash points, meaning they can ignite easily at relatively low temperatures. The volatility of flammable liquids makes them especially dangerous during storage and transportation, necessitating stringent handling procedures to prevent accidents. Accidental spills or leaks can lead to fires or explosions, thereby endangering lives and causing substantial property damage. Class 3 of hazardous materials classification specifically deals with flammable and combustible liquids, highlighting the critical nature of managing such materials safely (Environmental, 2017).

Explosives, such as fireworks, deserve special mention due to their capability to release energy rapidly, resulting in heat, pressure, sound, and light. The inherent dangers of explosives make them strictly regulated under Class 1 hazardous materials. Their handling requires specialized training to prevent accidental detonation, which can result in catastrophic outcomes. Even minor mishandling or environmental changes can trigger an explosive reaction, underscoring the need for comprehensive understanding and respect for these substances (<i>Nine Classes of Hazardous Materials (Yellow Visor Card) | FMCSA</i>, n.d.).

The potential hazards presented by these materials may manifest immediately or over long periods. Immediate hazards typically involve acute exposure where the effects are visible shortly after contact. For example, inhaling toxic fumes from a chemical spill can cause respiratory distress within minutes. Conversely, long-term hazards may emerge years after exposure, as seen with carcinogenic substances leading to cancer development. Therefore, understanding both short-term and long-term risks associated with hazardous materials is vital for implementing effective safety strategies.

Exposure routes are varied and can significantly influence the degree of risk posed by hazardous materials. Inhalation is a primary concern, especially for gases and volatile liquids. This route allows harmful substances to enter the respiratory system quickly, potentially leading to lung damage or systemic toxicity. Absorption through the skin is another common exposure pathway, particularly with corrosive substances that can damage tissues upon contact. Proper use of personal protective equipment (PPE), such as gloves and masks, is crucial to minimize exposure risks.

Classification of Hazardous Materials

Classifying hazardous materials is essential in managing and responding to any incidents involving these substances effectively. One of the primary systems for classification is by the Department of Transportation (DOT). DOT categorizes hazardous materials into nine distinct classes to ensure safe transportation and handling. Class 1 includes explosives, which are substances that can cause explosions due to chemical reactions. Class 2 covers gases, further divided into flammable gases like propane, non-flammable gases such as nitrogen, and toxic gases including chlorine. Other classes include flammable liquids (Class 3), flammable solids (Class 4), oxidizing substances and organic peroxides (Class 5), toxic and infectious substances (Class 6), radioactive material (Class 7), corrosive substances (Class 8), and miscellaneous dangerous goods (Class 9). Knowing these classifications helps responders anticipate the risks involved with specific materials and thus develop appropriate response strategies. Notably, the DOT system not only aids in identifying hazards but also influences packaging, labeling, and transportation procedures, ensuring uniformity across various sectors.

In addition to the DOT's systematic approach, another prominent method of classification is through the National Fire Protection Association's NFPA 704 system. Commonly known as the Fire Diamond, the NFPA 704 classifies materials based on health, flammability, reactivity, and special hazards using a color-coded symbol. Red indicates flammability, blue denotes health hazards, yellow represents reactivity, and white specifies special instructions or warnings such as unusual reactivity with water or the presence of an oxidizing agent (<i>EHS - NFPA Hazard Diamond | MyUSF</i>, n.d.). Each hazard is rated on a scale from 0 (no hazard) to 4 (severe hazard), allowing firefighters and emergency personnel to quickly assess the potential dangers at a glance. This rapid assessment capability is instrumental during emergencies, enabling responders to prioritize actions and deploy suitable countermeasures to mitigate risks. Unlike other systems, the NFPA 704 provides an immediate visual cue, making it indispensable in scenarios where time is of the essence.

The Globally Harmonized System (GHS) also plays a pivotal role in hazardous material classification. GHS is an internationally standardized system that enhances the consistency and clarity of hazard communication. Under GHS, chemical hazards are categorized into physical, health, and environmental hazards, each represented by specific pictograms, signal words (e.g., "danger" or "warning"), and precautionary statements. These elements appear on labels and Safety Data Sheets (SDS), offering comprehensive information about the potential dangers associated with a substance (Smith, 2015). By standardizing these elements worldwide, GHS facilitates international trade and improves safety in workplaces by ensuring that all parties have a unified understanding of material risks. For instance, the presence of a flame symbol universally indicates flammable material, allowing quick identification irrespective of language barriers.

Understanding these classification systems is critical for professionals dealing with hazardous materials. Proper classification guides the entire lifecycle of a hazardous material, affecting everything from storage, handling, and transportation to emergency response planning. When materials are correctly classified, responders can utilize appropriate protective equipment, evacuation plans, and neutralization methods. Furthermore, accurate classification ensures compliance with regulatory requirements, thereby influencing legal and insurance matters. It also minimizes the risk of accidents, as workers are better informed and prepared to manage potential hazards.

Moreover, integrated knowledge of these classification systems aids in developing effective response strategies. For example, knowing the differences between a flammable liquid and a flammable gas allows responders to choose the right fire suppression technique. While dry chemical extinguishers might be effective for liquid fires, gas fires may require a different approach, such as shutting off the gas supply. Similarly, understanding the significance of reactivity ratings helps in predicting the likelihood of chemical reactions that could exacerbate an emergency situation. In essence, comprehensive knowledge of these systems enables tailored responses that enhance safety and reduce potential harm during hazardous material incidents.

Importance of HazMat Training for Emergency Responders

In the realm of hazardous materials management, the necessity of HazMat training cannot be overstated. For emergency responders, such as firefighters and first responders, who are often the first on the scene during a hazardous incident, comprehensive training is pivotal. This specialized instruction equips them with the skills to operate safely and efficiently when dealing with dangerous substances, thereby safeguarding not only their own well-being but also that of the public.

Training programs play a crucial role in protecting both personnel and public safety during hazardous incidents. Without proper training, responders might lack the critical knowledge needed to manage these situations effectively, potentially leading to disastrous outcomes. For instance, trained individuals can quickly assess the situation, identify the materials involved, and implement appropriate control measures to prevent escalation. Additionally, they learn how to use personal protective equipment (PPE) correctly, which is indispensable in protecting themselves from harmful exposures.

Moreover, HazMat training ensures legal compliance and environmental protection, which are essential outcomes. Regulations governing hazardous materials are strict, and organizations must adhere to them to avoid legal repercussions. Proper training informs employees about these regulations and guidelines, ensuring they operate within the legal framework. Furthermore, understanding the potential environmental impact of hazardous materials allows trained personnel to take necessary precautions to prevent contamination and pollution.

An examination of past industrial disasters underscores the importance of adequate HazMat training. Analyzing case studies such as the Bhopal Gas Tragedy and the Chernobyl Nuclear Disaster offers invaluable lessons. The Bhopal disaster occurred in December 1984 when a pesticide plant in India leaked methyl isocyanate gas, resulting in thousands of casualties and long-term health issues for survivors. This tragedy highlighted serious deficiencies in safety protocols and emergency response planning. Similarly, the catastrophic events at Chernobyl in 1986 revealed the grave consequences of inadequate training and oversight in nuclear facilities. Both incidents demonstrate how comprehensive training could have mitigated the severity of these disasters by preparing workers to respond more effectively.

Continuous education through drills is another vital component of HazMat training, ensuring readiness and knowledge enhancement among personnel. Regular drills mimic real-life situations, allowing responders to practice their skills in a controlled environment. These exercises help reinforce theoretical knowledge, build confidence, and identify any weaknesses in current procedures. By conducting consistent drills, organizations maintain a high state of preparedness, ensuring that responders remain alert and capable when faced with actual emergencies.

Guidelines for continuous education emphasize the importance of integrating updated information into training modules. As new technologies emerge and industry standards evolve, training programs must adapt accordingly to include this latest knowledge. Furthermore, fostering an environment where questions are encouraged and discussions are part of the learning process enhances the overall educational experience for participants.

The significance of HazMat training extends beyond immediate crisis management. It fosters a culture of safety within organizations, encouraging a proactive approach to hazard identification and risk mitigation. When workers are aware of the potential risks and are trained to handle them adeptly, they contribute actively to maintaining a safe workplace.

Finally, considering the diverse and ever-changing landscape of hazardous materials, ongoing education is key. The dynamic nature of industries dealing with such substances calls for a constant update of skills and knowledge. Regular refreshers and advanced courses ensure that all personnel, regardless of their level of experience, are equipped to face new challenges.

Laws and Regulations Governing Hazardous Materials

The regulatory framework governing hazardous materials is essential in safeguarding public health and the environment. A crucial component of this framework is OSHA's Hazardous Waste Operations and Emergency Response (HAZWOPER) Standard. This guideline requires the use of Personal Protective Equipment (PPE) and focuses on worker safety during hazardous material responses. It establishes protocols to ensure workers are adequately trained and protected from potential harm when handling hazardous substances. The HAZWOPER standard, documented under 29 CFR 1910.120, requires employers engaged in hazardous waste operations to develop comprehensive safety programs, conduct regular training, and implement medical surveillance for individuals exposed to hazardous substances. These measures significantly contribute to minimizing risks such as chemical burns, toxic exposure, and other health hazards. Compliance with HAZWOPER also emphasizes the necessity of annual refresher training, ensuring that workers remain informed about proper emergency response practices (<i>Hazardous Waste

Operations and Emergency Response (HAZWOPER) - Standards | Occupational Safety and Health Administration</i>, 2025).

In addition to OSHA regulations, the Department of Transportation (DOT) plays a pivotal role in managing the safe transportation of hazardous materials. Governed by the Hazardous Materials Transportation Act, DOT regulations are designed to protect the public and environment from incidents involving hazardous material transport. These regulations cover packaging, labeling, and documentation to ensure proper handling during transit. For instance, hazardous materials must be securely packaged to prevent leaks or spills, and clearly labeled to communicate their contents and associated dangers. Through rigorous compliance measures, DOT regulations aim to prevent accidents that could lead to injury or environmental damage. By enforcing these standards, the DOT ensures that individuals involved in the transportation process are aware of potential hazards and equipped to manage them safely.

Equally important within the regulatory framework is the Environmental Protection Agency's (EPA) Superfund Amendments and Reauthorization Act (SARA) Title III, which reinforces the community's right-to-know about chemical hazards. This provision mandates that facilities using or producing hazardous chemicals provide detailed information to local governments and the public. By doing so, SARA Title III empowers communities to effectively prepare for and respond to chemical emergencies. Information disclosed under this act includes safety data sheets, chemical inventories, and release reporting requirements. Such transparency not only fosters public trust but also enables communities to take proactive measures in protecting themselves and the environment from potential chemical risks. This regulation illustrates how providing accurate and timely information can enhance collective safety and preparedness.

Compliance with National Fire Protection Association (NFPA) standards further complements these regulatory efforts by ensuring professional competency and accuracy in labeling. NFPA standards encompass various aspects of fire safety and hazard communication crucial for handling hazardous materials. Adherence to these standards guarantees that professionals possess the necessary skills and knowledge to manage hazardous situations effectively. Moreover, NFPA's provisions on labeling facilitate clear identification of hazardous materials, thereby aiding in swift and accurate emergency responses. Labels often use standardized symbols and colors to convey critical information such as flammability, reactivity, and specific precautions. This uniformity assists first responders, firefighters, and other stakeholders in quickly assessing risks and taking appropriate actions to mitigate them.

Concluding Thoughts

In this chapter, we explored the fundamental aspects of hazardous materials, focusing on how they are classified and the significance of proper training and regulatory compliance. We examined the various types of hazardous materials such as toxic substances, flammable liquids, and explosives, emphasizing their potential risks to life, health, property, and the environment. Recognizing these dangers is critical for those who handle them daily, including students, emergency responders, and professionals. We also discussed key classification systems like the Department of Transportation (DOT), NFPA 704, and the Globally Harmonized System (GHS), highlighting their roles in managing the safe storage, handling, and transportation of hazardous substances.

Understanding hazardous materials extends beyond academic knowledge; it requires practical training and adherence to relevant laws and regulations. The importance of HazMat training was underscored through real-world examples, demonstrating its vital role in preparing responders to manage emergencies effectively. Additionally, we touched on the crucial guidelines set by OSHA, DOT, and EPA, which safeguard public health and the environment. By comprehending these classifications and adhering to legal requirements, individuals can enhance their safety strategies, ensure compliance, and protect communities from potential chemical hazards. As you move forward, keep these insights in mind to better prepare for the challenges associated with hazardous materials management.

20 questions and answers based on the chapter 1"Introduction to Hazardous Materials":

What are hazardous materials?

Answer: Hazardous materials are substances that pose significant risks to life, health, property, and the environment due to their chemical and physical properties. They can cause harm through inhalation, skin absorption, ingestion, or direct contact.

What is the difference between hazardous materials and hazardous waste?

Answer: Hazardous materials are substances that are actively used or transported and pose immediate threats, while hazardous waste refers to substances that are disposed of after use but still have the potential to harm the environment or health if not managed properly.

What are some examples of hazardous materials?

Answer: Examples include toxic substances (like lead and mercury), flammable liquids (like gasoline), and explosives (like fireworks).

How do hazardous materials present risks?

Answer: They can present immediate hazards through acute exposure, such as inhalation of toxic fumes, or long-term hazards like carcinogenic substances leading to cancer.

What is the importance of classifying hazardous materials?

Answer: Classification helps in managing hazardous materials by ensuring proper handling, packaging, labeling, and emergency response. It guides responders on the risks involved and how to address them.

What are the nine classes of hazardous materials according to the Department of Transportation (DOT)?

Answer: The nine classes are:
Explosives
Gases
Flammable liquids
Flammable solids
Oxidizing substances and organic peroxides
Toxic and infectious substances
Radioactive materials
Corrosive substances
Miscellaneous dangerous goods.

What does the NFPA 704 Fire Diamond indicate?

Answer: The NFPA 704 Fire Diamond uses a color-coded system to indicate the severity of hazards: red for flammability, blue for health hazards, yellow for reactivity, and white for special hazards.

What is the Globally Harmonized System (GHS)?

Answer: GHS is an internationally standardized system that classifies chemical hazards into physical, health, and environmental categories. It uses specific pictograms, signal words, and precautionary statements on labels and Safety Data Sheets (SDS) to convey these hazards.

Why is HazMat training important for emergency responders?

Answers: HazMat training equips responders with the skills needed to handle hazardous materials safely, protecting both themselves and the public. It ensures proper use of personal protective equipment (PPE) and compliance with safety regulations.

What role does HazMat training play in legal compliance?

Answer: HazMat training ensures that responders are aware of and follow the legal guidelines and regulations set by OSHA, DOT, and EPA, minimizing the risk of legal consequences and environmental harm.

What was the Bhopal Gas Tragedy, and what lesson does it teach?

Answer: The Bhopal Gas Tragedy in 1984 involved a toxic gas leak that caused thousands of deaths. It highlights the need for effective training and safety protocols to prevent such disasters.

What happened during the Chernobyl Nuclear Disaster?

Answer: The Chernobyl disaster in 1986 involved a nuclear meltdown that resulted in widespread radiation exposure. It emphasizes the importance of adequate training in managing hazardous materials in nuclear facilities.

What is the role of drills in HazMat training?

Answer: Drills simulate real-life situations and allow responders to practice their skills, reinforce theoretical knowledge, and identify weaknesses in procedures, ensuring they are prepared for actual emergencies.

What is the HAZWOPER standard, and how does it impact HazMat training?

Answer: The HAZWOPER standard, enforced by OSHA, requires employers to develop safety programs, provide regular training, and use PPE for workers involved in hazardous waste operations, ensuring their protection during emergencies.

How does the DOT regulate the transportation of hazardous materials?

Answer: The DOT ensures the safe transportation of hazardous materials through packaging, labeling, and documentation requirements to prevent leaks, spills, and accidents during transit.

What is the purpose of the EPA's SARA Title III?

Answer: SARA Title III mandates that facilities using hazardous chemicals provide information to local governments and the public, allowing communities to prepare for and respond to chemical emergencies.

What is the significance of NFPA standards in HazMat management?

Answer: NFPA standards ensure that professionals are properly trained and equipped to handle hazardous materials, providing clear labeling and guidelines for safety during hazardous material incidents.

Why is it essential to have accurate classification of hazardous materials?

Answer: Proper classification allows responders to determine appropriate safety measures, response strategies, and protective equipment, ensuring effective management and reducing risks during hazardous material incidents.

How do exposure routes affect the severity of hazards?

Answer: Exposure routes like inhalation, skin absorption, and ingestion determine how harmful a hazardous material can be. Some routes, such as inhalation, can lead to immediate effects, while others, like skin absorption, can cause long-term damage.

What are the main hazards posed by flammable liquids?

Answer: Flammable liquids, like gasoline, can ignite easily at low temperatures, presenting fire and explosion risks during storage, transportation, or accidental spills. Proper handling and safety measures are essential to prevent such accidents.

Chapter 2: Hazard Recognition and Identification

(Awareness Level)

Hazard recognition and identification at the awareness level serves as a foundational skill set for those encountering potential hazards, including hazardous materials students, firefighters, and first responders. Understanding these concepts not only promotes personal safety but also establishes a framework for effective and efficient emergency response. By recognizing hazardous situations early on, individuals can take decisive action that mitigates risks and protects both individuals and communities from harm. This awareness becomes crucial in various environments where hazardous materials may be present, offering a proactive approach to manage and contain dangers before they escalate into emergencies.

This chapter delves into the importance of hazard recognition and identification by examining a range of real-world contexts and scenarios. It explores how visual indicators like warning signs and placards serve as essential tools for immediate hazard identification. Additionally, the role of sensory cues, such as odors or sounds, is discussed in relation to enhancing situational awareness. Beyond this, the chapter highlights the significance of documentation, including Safety Data Sheets (SDS), in providing critical information about hazardous substances. Furthermore, technological advancements that aid in hazard detection are examined to show how they contribute to safer operational practices. Engaging workers in hazard recognition and selecting appropriate controls form key discussions to emphasize the collaborative nature of maintaining safety. Overall, this chapter equips readers with the knowledge and strategies necessary for robust hazard identification and management, ensuring compliance with regulations and fostering a culture of safety.

Introduction to Hazard Recognition and Identification

Recognizing and identifying hazardous materials is critical for safeguarding the well-being of first responders and the public. The ability to promptly identify a hazard ensures that emergency actions can be taken before situations escalate, reducing potential harm and preventing further risks. For instance, quick identification of a chemical spill can allow for faster containment and evacuation procedures, significantly limiting exposure to dangerous substances.

Effective hazard recognition also plays a vital role in maintaining compliance with safety regulations. Adherence to these regulations is not only a legal obligation but also a moral duty to ensure the safety of all individuals involved in or near hazardous situations. By following established guidelines and protocols, organizations create a safe operational environment, minimizing liabilities and enhancing trust with stakeholders.

To achieve effective hazard recognition, various methods should be employed. Visual indicators, such as warning signs and labels, are crucial tools in identifying hazardous materials. These visual cues provide immediate information about the potential dangers present, allowing first responders to act accordingly. For example, the use of standardized labels on chemical containers can quickly inform personnel of the specific risks involved, aiding in the decision-making process during an emergency.

Sensory cues can also indicate the presence of hazardous materials. Odors, sounds, and even changes in ambient temperature can signal potential dangers. First responders trained to detect these sensory warnings can often recognize hazards before they become immediately visible, providing a critical window of opportunity for intervention. An understanding of these sensory cues allows for preemptive measures, ensuring the safety of personnel and the public.

Documentation plays a significant role in hazard recognition as well. Proper records, including safety data sheets (SDS) and incident reports, offer detailed information regarding the properties and handling procedures of hazardous materials. Accurate documentation allows personnel to familiarize themselves with potential hazards in advance, facilitating informed responses during emergencies. By incorporating comprehensive record-keeping practices, organizations bolster their preparedness for hazardous situations.

The integration of technology into hazard recognition efforts further enhances safety measures. Advanced detection systems, such as gas detectors and radiation sensors, provide real-time data on environmental conditions, alerting first responders to potential threats. Technologies like the Social Media Analytics and Reporting Toolkit (SMART) enhance situational awareness by collecting and analyzing social media data for actionable insights, helping in the swift recognition of emerging hazards (<i>First Responder Technologies | Homeland Security</i>, 2023).

Furthermore, the application of Precision Outdoor and Indoor Navigation and Tracking (POINTER) technology allows for accurate positioning of first responders in complex environments. By maintaining awareness of responder locations, POINTER enhances response efforts and ensures the timely identification of lost or injured personnel, thus improving overall safety outcomes (<i>First Responder Technologies | Homeland Security</i>, 2023).

Employers must actively involve workers in the hazard recognition process. Workers often possess invaluable knowledge of facility conditions and potential hazards. Their insights can help in evaluating the feasibility and effectiveness of control measures. Engaging workers in discussions on hazard identification fosters a culture of safety and encourages collaborative problem-solving, ultimately leading to more robust safety practices (Osha, 2024).

Selecting appropriate controls is another essential aspect of hazard management. Employers are encouraged to eliminate or control serious hazards immediately, using a hierarchy of controls that prioritizes engineering solutions, safe work practices, administrative controls, and personal protective equipment. Effective selection of controls reduces the likelihood of escalating hazards and safeguards the health of both responders and the public (Osha, 2024).

In addition, developing and updating a hazard control plan ensures that selected controls are implemented efficiently. An effective plan addresses serious hazards and outlines interim controls for ongoing protection while long-term solutions are developed. Regular updates and evaluations of the control plan guarantee the continued efficacy of safety measures, adapting to changing conditions and technologies (Osha, 2024).

Moreover, selecting controls to protect workers during nonroutine operations and emergencies is crucial. Conducting job hazard analyses and notifying workers of potential risks before undertaking unusual tasks minimizes the chances of unforeseen accidents. Comprehensive planning and communication mitigate risks associated with unpredictable scenarios, enabling first responders to manage emergencies with confidence (Osha, 2024).

Recognizing Hazardous Materials in Various Contexts

Hazardous materials (HazMat) are a crucial consideration across various settings due to their potential risks to health, safety, and the environment. Recognizing and identifying these materials in different contexts is vital for effective management, response, and prevention of hazardous incidents. Understanding where these materials can be found allows first responders, firefighters, and students of hazardous materials to better prepare and respond effectively.

In the realm of transportation, hazardous materials are frequently encountered. These materials are discernable through specific transport vehicle types and Department of Transportation (DOT) markings. For instance, tank cars and multi-unit tank car tanks are common in rail transport and are easily identifiable by their structure. DOT placards and labels play a critical role in this setting. Each bulk packaging, freight container, or transport vehicle containing hazardous materials must display the appropriate placards on each side and end. These placards include visual indicators such as colors and symbols that communicate the type of hazard present, such as flammable liquids or corrosive substances (FMCSA, 2013). Proper labeling and marking ensure that those handling the cargo can identify hazards quickly and respond appropriately.

Industrial and commercial settings are another hotspot for hazardous materials. Chemical plants, refineries, and mixed storage warehouses often house significant quantities of chemicals that could pose substantial risks if mishandled. Within these environments, hazardous materials are typically stored in large containers, drums, or specialized facilities designed for safe containment. Additionally, these sites employ rigorous protocols to manage and mitigate risks associated with chemical handling and storage. In these industrial settings, knowledge of labeling requirements and the placement of safety signs contributes to maintaining a safe operational environment (US EPA, 2019).

In everyday environments, the presence of hazardous materials is often underestimated. Household products like cleaning agents, paints, and pesticides contain chemicals that, while useful, can be dangerous if misused or spilled. Likewise, agricultural chemicals such as fertilizers and insecticides are common in farming communities and require careful handling and storage. Medical biohazards, particularly in healthcare settings, also present unique challenges. Items like sharps, biological waste, and chemicals used in medical research necessitate strict adherence to safety protocols to prevent exposure and contamination. Recognizing hazards in these everyday environments is essential for preventing accidents and safeguarding public health. Guidelines about safely storing and disposing of household chemicals should be followed to minimize risks.

Documentation serves as an indispensable asset for identifying hazardous materials across all contexts. Shipping documents, including Bills of Lading and Safety Data Sheets (SDS), provide comprehensive information about the materials being transported or stored. These documents include details such as the proper shipping name, identification numbers, and emergency response information. For transportation specifically, the shipping paper accompanying hazardous waste is crucial; it ensures that responders have immediate access to necessary details for handling materials safely (US EPA, 2019). Having accurate documentation enables workers and first responders to understand the nature of the hazards they are dealing with, thereby enhancing safety measures.

Importantly, the systematic approach to identifying hazardous materials through transport markings, industrial protocols, consumer awareness in everyday environments, and thorough documentation underscores the integrated strategy required to manage HazMat effectively. By adhering to established guidelines and regulations, individuals and organizations can significantly reduce the risk of hazardous incidents. For instance, ensuring that emergency contacts are accurate and accessible on shipping documents facilitates prompt and informed responses during emergencies.

DOT Hazard Classes and Labeling Systems

The Department of Transportation (DOT) classification and labeling system is a cornerstone in ensuring the safe transport and handling of hazardous materials. Understanding this system is essential for first responders, firefighters, and hazardous materials students to make informed decisions during emergencies.

Categorization of materials based on risk is the first step in comprehending the potential dangers associated with each substance. The DOT has organized hazardous materials into classes that reflect their specific hazards, such as flammability, corrosiveness, or toxicity. For instance, Class 1 encompasses explosives, which are further divided into subdivisions based on their explosion and projection hazards. These divisions include powerful mass explosion hazards under Division 1.1 and less severe threats like minor blast hazards under Division 1.4 (<i>DOT Hazard Classes</i>, n.d.). This systematic categorization allows for quick identification of hazards and aids in determining appropriate response strategies, thereby mitigating risks effectively.

Proper labeling plays a vital role in enabling emergency responders to act swiftly and safely. Labels provide crucial information at a glance, allowing responders to assess the situation quickly and decide on the necessary precautions and actions. When responding to incidents involving hazardous materials, time is of the essence, and proper labeling facilitates rapid decision-making, reducing the potential for harm to both responders and the public. It is through these labels that responders can identify the class of substance they are dealing with and apply the pertinent safety measures.

Each class of hazardous material, from explosives to miscellaneous goods, possesses unique characteristics. For example, Class 2 deals with gases, including flammable gases like propane under Division 2.1 and non-flammable, non-toxic gases such as nitrogen under Division 2.2 (<i>DOT Hazard Classes</i>, n.d.). These distinctions help responders understand the nature of the hazard and the appropriate containment or mitigation tactics. Similarly, Class 3 focuses on flammable liquids, where identifying whether a liquid is merely flammable or also combustible can alter the approach to managing a spill or leak.

An essential component of the DOT system is its placard requirements, which utilize a clear color-coded and symbol-based framework. Placards are required when transporting hazardous materials and must display the hazard class or division number prominently. The utilization of color coding and symbols means that even from a distance, responders can determine the type of hazard they are encountering. For example, a red background universally indicates flammable materials, while yellow signifies an oxidizing agent, and white with black symbols denotes radioactive substances (<i>DOT Hazard Classes</i>, n.d.).

These placards not only comply with national regulations but also align with international standards, promoting consistent communication of hazards across borders. This consistency ensures that no matter where an incident occurs, trained individuals can respond appropriately, leveraging the universal language of the placard system.

While navigating the DOT classification and labeling system may initially seem daunting due to its complexity, it is invaluable for effective hazard management and emergency response. By internalizing the categories and corresponding labels, responders are better equipped to protect themselves, others, and the environment from the dangers posed by hazardous materials. Through training and practice, first responders can become adept at interpreting these labels swiftly, minimizing reaction times in critical situations.

Moreover, the importance of having comprehensive guidelines for interpreting NFPA labels in emergency situations cannot be overstated. Such guidelines should be accessible and straightforward, providing a step-by-step approach for responders who may face high-pressure scenarios where clear, decisive action is needed.

NFPA 704 and SDS Utilization

In the realm of hazard management, understanding and using tools like the NFPA 704 system and Safety Data Sheets (SDS) is essential for effective safety practices. These resources provide critical information about hazardous materials that can help in both immediate risk assessment and long-term planning, particularly for those in professions handling such materials regularly.

The NFPA 704 system, commonly recognized as the "NFPA Diamond," serves as a quick reference tool for emergency responders to identify potential risks associated with various hazardous materials during an incident. Each segment of the diamond uses specific colors and numerical ratings to indicate the level of danger in four key categories: health, flammability, reactivity, and special hazards. This enables responders to gauge the severity of the risks they might face swiftly. The color coding is quite intuitive—blue represents health hazards, red is for flammability, yellow denotes reactivity, and white is reserved for any special notices related to the material's properties. For instance, a higher number in the flammability section implies a greater risk of ignition under normal use conditions. Understanding this system not only aids first responders but also enhances the situational awareness of all personnel involved in handling or responding to hazardous materials. It is crucial for everyone involved to be familiar with the meaning behind each color and rating, as it dictates the precautions and response measures required in case of exposure (<i>NFPA 704 Symbols - Institutional Risk & Safety | the University of Texas at Dallas</i>, n.d.).

Complementing the NFPA 704 system are Safety Data Sheets (SDS), which provide a more detailed overview of hazardous substances. These documents are indispensable as they offer comprehensive information on how to handle, store, and dispose of chemical materials safely. Structured according to international standards, SDS include sections detailing everything from potential hazards and first-aid measures to proper storage practices and personal protective equipment recommendations. Sections specifically relevant to hazard identification align with some aspects of the NFPA 704, such as health and flammability, further reinforcing the understanding necessary for safe handling. By thoroughly reviewing these sheets, individuals can ensure compliance with regulatory requirements while effectively managing emergencies that involve chemical exposures. Moreover, when accessed quickly during a crisis, SDS can guide the development of a targeted emergency response plan that minimizes risks to both people and the environment (<i>EHS - NFPA Hazard Diamond | MyUSF</i>, n.d.).

A critical aspect of using SDS effectively is knowing how to access them promptly and interpret their contents accurately. In an emergency, time is often a limited resource. Therefore, having processes in place that allow for rapid retrieval of these documents is vital. Many organizations maintain digital SDS repositories that can be accessed via computers or mobile devices, ensuring that even when physical copies are unavailable, essential data remains at one's fingertips. Interpreting these documents demands familiarity with their structure and terminology; therefore, ongoing training and drills should be conducted to keep staff proficient in utilizing SDS during real-world scenarios. Such preparation allows personnel to swiftly enact response measures tailored to the specific hazards identified, thereby reducing potential harm and facilitating a more controlled resolution of incidents.

Utilizing both the NFPA 704 system and SDS provides a comprehensive approach to hazard management. While the NFPA diamond offers an immediate visual summary of a material's dangers—ideal for making quick decisions—the SDS delivers an in-depth perspective needed for crafting thorough response strategies and ensuring ongoing compliance with safety regulations. Together, they form the cornerstone of a robust safety program that prioritizes the well-being of workers and the community. As highlighted by both sources, understanding these tools is fundamental not just for those in firefighting or emergency roles but also for anyone who engages with potentially hazardous materials in their day-to-day operations.

By integrating these resources into standard operating procedures, organizations can significantly enhance their preparedness for dealing with hazardous material incidents. This proactive approach ensures that staff are equipped not only with the knowledge to recognize warning signs swiftly but also with the information needed to respond efficiently. The synergy between the NFPA 704 system and SDS exemplifies how structured information can empower individuals to manage risks effectively, ultimately safeguarding lives and property.

Core Message

Understanding and identifying hazardous materials is a vital skill for safeguarding the safety of first responders, the public, and overall community well-being. Throughout this chapter, we explored the importance of recognizing hazards promptly to enable effective emergency responses before situations escalate. We delved into various methods for hazard recognition, including visual and sensory cues, the significance of proper documentation, and the integration of advanced technology like POINTER and SMART systems. By harnessing these resources, individuals responsible for handling hazardous materials can enhance their situational awareness, enabling quicker and more informed decision-making during emergencies.

Additionally, we highlighted the critical role of compliance with safety regulations in maintaining a safe environment. This involves engaging workers actively in hazard recognition processes and implementing effective control measures to manage risks associated with nonroutine operations and emergencies. Utilizing structured tools such as DOT hazard classes, NFPA 704 labels, and Safety Data Sheets was emphasized as indispensable for conveying crucial hazard information efficiently. The comprehensive strategies discussed underscore the need for ongoing training and collaboration among all involved parties, ensuring that they are equipped with the knowledge necessary to address potential risks effectively and protect human health and safety.

20 questions with answers based on Chapter 2: Hazard Recognition and Identification (Awareness Level):

What is the importance of hazard recognition and identification at the awareness level?

Answer: It is critical for safeguarding the well-being of first responders and the public by enabling early recognition of hazards, allowing for timely and effective emergency responses.

What are some common visual indicators used for hazard recognition?

Answer: Warning signs, placards, labels, and color codes are commonly used to visually identify hazardous materials and indicate potential risks.

What sensory cues can aid in hazard recognition?

Answer: Sensory cues such as odors, sounds, and temperature changes can indicate the presence of hazardous materials before they are visibly noticeable.

How does documentation support hazard recognition?

Answer: Documentation like Safety Data Sheets (SDS) and incident reports provide essential information about hazardous substances, aiding in understanding their properties and safe handling procedures.

What role does technology play in hazard recognition?

Answer: Technologies like gas detectors, radiation sensors, and situational awareness tools like SMART systems help in detecting and alerting responders to environmental hazards in real-time.

How do warning signs and placards help first responders?

Answer: They provide immediate visual information about the type of hazard present, helping first responders quickly assess the situation and take appropriate action.

What is the role of workers in hazard recognition?

Answer: Workers have valuable insights into potential hazards in their environments and are crucial in evaluating risks and control measures, helping to create a safer workplace.

What is the hierarchy of controls in hazard management?

Answer: The hierarchy of controls prioritizes eliminating hazards, followed by engineering controls, safe work practices, administrative controls, and personal protective equipment.

Why is developing and updating a hazard control plan important?

Answer: It ensures that selected control measures are implemented effectively and that interim solutions are in place while long-term measures are developed and refined.

How does hazard recognition apply in transportation settings?

Answer: Hazardous materials are often transported in specific vehicles (e.g., tank cars) and are marked with placards and labels to indicate the type of hazard, enabling responders to act quickly.

What types of hazardous materials are commonly found in industrial settings?

Answer: Industrial settings, such as chemical plants and refineries, house hazardous chemicals in large containers or specialized storage, requiring strict safety protocols to prevent risks.

How are hazardous materials identified in everyday environments?

Answer: Household products like cleaning agents and pesticides, as well as agricultural chemicals, contain hazardous substances that require careful handling and storage.

What role does shipping documentation play in hazard recognition?

Answer: Shipping papers, including Bills of Lading and Safety Data Sheets (SDS), provide critical information about the hazardous materials being transported, ensuring safe handling and emergency response.

How does the DOT classification system aid in hazard recognition?

Answer: The DOT system categorizes hazardous materials into classes based on their specific risks (e.g., flammable liquids, explosives) and provides corresponding placards to facilitate quick identification.

What is the NFPA 704 system, and how is it used?

Answer: The NFPA 704 system, or "NFPA Diamond," uses color-coded segments to indicate health, flammability, reactivity, and special hazards, helping responders assess risks quickly.

What information is provided in a Safety Data Sheet (SDS)?

Answer: SDS provide detailed information about the hazards of chemicals, including handling procedures, first-aid measures, storage instructions, and recommendations for personal protective equipment.

Why is it important to understand and use the NFPA 704 system and SDS during emergencies?

Answer: They provide critical information needed for quick risk assessment and safe handling during emergencies, improving responders' ability to make informed decisions.

What is the role of placards in the DOT system?

Answer: Placards are used to display hazard class information on transport vehicles, allowing responders to quickly determine the type of hazard they are dealing with from a distance.

How can technological systems like POINTER improve hazard recognition?

Answer: POINTER technology provides real-time tracking of first responders in complex environments, enhancing safety and aiding in the identification of hazards or missing personnel.

How can engaging workers in hazard recognition contribute to a culture of safety?

Answer: Involving workers in hazard recognition fosters collaboration, encourages proactive problem-solving, and strengthens safety practices by utilizing their knowledge of facility conditions and potential risks.

Chapter 3: Initial Actions and Scene Management

(Awareness Level)

Managing hazardous material (HazMat) incidents at the awareness level requires a clear understanding of initial actions and effective scene management. These specialized tasks are pivotal for ensuring the safety and well-being of both responders and the public. As the first line of response, those at the awareness level play an essential role in assessing threats and implementing critical measures to control dangerous situations. This involves making informed decisions quickly and with precision, setting the stage for orderly, strategic interventions. Each response begins with a methodology that prioritizes containment strategies, evaluates risks, and coordinates efforts among various stakeholders. Proper management techniques help prevent escalation, reducing the potential for harm and ensuring public safety. Given these responsibilities, a systematic approach is non-negotiable for maintaining control and ensuring efficient resource deployment.

In this chapter, we will explore the key steps and responsibilities necessary for managing HazMat incidents effectively at the awareness level. The discussion highlights the importance of coordinated efforts and strategic decision-making in protecting people and the environment. Key topics include protocols for controlling exposure to hazardous substances, establishing isolation zones, and optimizing communication channels. We will also examine the role of training in preparing responders to handle emergencies with confidence and skill. Additionally, the chapter delves into the creation and maintenance of isolation zones, such as Hot, Warm, and Cold zones, to ensure containment and proper execution of rescue and cleanup operations.

Lastly, it addresses the necessity of adhering to standardized protocols that guide responders through every stage of an incident, ensuring compliance with regulations and promoting seamless coordination across different agencies. With a focus on structured operations and clear communication, this chapter aims to equip novices and seasoned responders alike with the knowledge needed to manage scenes efficiently and safely.

Importance of Proper Scene Management

Responding effectively to hazardous material (HazMat) incidents is a crucial responsibility for emergency personnel, particularly at the awareness level. One of the most important aspects of handling such incidents is effective scene management. This involves coordinated efforts and strategic decision-making that ensure the safety of individuals and the environment. Awareness-level responders, often the first to arrive at the scene, must understand how proper scene management can significantly reduce risks associated with HazMat incidents.

Effective scene management begins with controlling exposure to hazardous materials. Responders play a critical role in assessing the potential dangers posed by a spill or chemical release. By quickly identifying the type and magnitude of the hazard, they are better equipped to implement strategies that protect both themselves and the public from exposure. For instance, responders might work to contain a chemical spill to prevent it from spreading to populated areas or waterways, thus reducing potential health risks and environmental damage. The ability to control exposure not only enhances safety but also limits the area that needs to be evacuated or cordoned off, minimizing disruption (<i>The Golden First Minutes — Initial Response to a Chemical Hazardous Materials Incident - CHEMM</i>, n.d).

Moreover, effective scene management prevents escalation of incidents. Inadequately managed scenes can rapidly spiral into more serious situations that threaten lives and infrastructure. HazMat incidents have the potential to cause fires, explosions, or toxic releases if not handled properly. By maintaining control over the incident scene, responders can stabilize the situation, preventing small-scale hazards from becoming large-scale disasters. Implementing containment measures swiftly and efficiently helps avoid further complications, ensuring that the impact on the community and environment is minimized (<i>Traffic Incident Management in Hazardous Materials Spills in Incident Clearance: 1.0 Introduction - FHWA Office of Operations</i>, 2025).

A standardized protocol is essential for effective scene management during HazMat incidents. Adhering to established guidelines ensures that all actions taken are consistent with best practices in emergency response. These protocols serve as a blueprint for responders, guiding them through each step of managing the scene—from initial assessment to requesting additional resources. Following these protocols not only enhances the safety and efficiency of operations but also ensures compliance with legal and regulatory requirements governing hazardous materials. It forms the basis for coordinated efforts among different agencies involved in the response process, including fire departments, law enforcement, and environmental protection organizations (<i>The Golden First Minutes — Initial Response to a Chemical Hazardous Materials Incident - CHEMM</i>, n.d).

Adopting a systematic approach to scene management fosters safer operations by promoting clear communication among responders. This includes keeping lines of communication open with incident commanders and other key personnel who may join later. Regular updates about the status of the incident and any changes in conditions enable quick adaptations to the strategy deployed, allowing for a more dynamic and responsive management style. Effective communication is also vital for informing the public and other stakeholders about necessary precautions and ongoing efforts to resolve the situation safely (<i>Traffic Incident Management in Hazardous Materials Spills in Incident Clearance: 1.0 Introduction - FHWA Office of Operations</i>, 2025).

Training plays an indispensable role in equipping responders with the skills required for managing HazMat scenes effectively. Comprehensive training programs should focus on recognizing various types of hazards, understanding their implications, and deploying appropriate measures to mitigate risks. Training also instills confidence in responders to make decisions under pressure, a critical component of successful scene management. Furthermore, ongoing drills and exercises can help reinforce this training, allowing responders to practice and refine their skills in a controlled environment. The goal is to ensure that when faced with real-world scenarios, responders can apply their training seamlessly, leading to more effective scene management and improved outcomes (<i>The Golden First Minutes — Initial Response to a Chemical Hazardous Materials Incident - CHEMM</i>, n.d).

Establishing Scene Safety and Isolation Zones

Defining safe operational areas during a hazardous material (HazMat) incident is crucial for ensuring the safety of both responders and the public. The process begins with identifying potential hazards, which involves assessing risks associated with fire, explosions, and toxic releases. These assessments are essential because they determine how threats can affect human health and safety. For instance, certain chemicals may pose higher risks under specific conditions, such as when exposed to heat or moisture. Understanding these variables helps responders establish appropriate precautionary measures promptly.

One of the primary methods for managing these risks is through setting up isolation zones, commonly referred to as Hot, Warm, and Cold zones. Establishing these zones aids in controlling contamination spread and organizing responder activities efficiently. The Hot Zone, or Exclusion Zone, is the area with actual or potential contamination and presents the highest risk of exposure to hazardous substances (US EPA, OLEM, 2013). Only personnel equipped with proper protective gear should enter this zone, and their tasks might include containment efforts or lifesaving rescues.

Adjacent to the Hot Zone is the Warm Zone, also known as the Contamination Reduction Zone. This transitional area serves as a buffer between contaminated and uncontaminated environments. It is where decontamination processes occur to ensure that hazardous substances do not migrate beyond this point. Activities within the Warm Zone require careful planning, including the setup of decontamination stations and the monitoring of contamination levels on equipment and personal protective clothing.

The final boundary is the Cold Zone, or Support Zone, which remains free from contamination (US EPA, OLEM, 2013). This zone functions as a command and control center where logistical operations, medical support, and other planning activities take place. Establishing these zones not only minimizes contamination but also organizes teams so they can perform specialized tasks without cross-contaminating efforts.

Utilizing the Emergency Response Guidebook (ERG) is another pivotal step in defining isolation distances. The ERG assists responders by providing guidelines on initial isolation and protective action distances for various chemicals. Depending on factors like spill size and weather conditions, the guidebook offers detailed tables specifying how far away people need to be evacuated to stay safe (<i>The First Responder: Technically Speaking</i>, 2025).

For example, a small chemical spill confined to less than 55 gallons will have different isolation requirements compared to a larger spill exceeding that volume. Weather conditions such as wind speed and direction play a significant role in determining how quickly harmful substances disperse. During inclement weather, especially strong winds, the potential spread of airborne toxins increases significantly, necessitating wider evacuation zones.

By interpreting data from the ERG, responders can accurately assess each incident's specific needs. Calculations for isolation and protective action distances hinge on several factors including substance toxicity, release method, and environmental interactions. This analytical framework allows emergency personnel to make informed decisions swiftly, reducing the likelihood of exposure and subsequent harm.

Furthermore, it's vital to adapt zoning procedures based on real-time observations and updates from ongoing assessments. As new information becomes available during an incident, adjustments to zone boundaries or protective actions might be warranted. Safety officers should be vigilant in monitoring environmental changes and disseminate updated instructions to all response teams effectively.

Given the dynamic nature of HazMat emergencies, clear communication lines are paramount for coordination. Any changes in zone specifications must be relayed promptly to avoid confusion among team members operating at different site levels. Utilizing technology such as mobile devices or radios ensures continuous contact among squads, improving operational efficiency under pressure.

While these guidelines provide an overarching framework, the nuances of each incident require tailored responses. Training scenarios often incorporate practical exercises simulating a range of emergency conditions. Such drills prepare teams to apply learned principles in high-stakes situations, enhancing their competency and confidence during real-world interventions.

Effective Reporting and Notifications

As first responders and hazardous materials students are aware, effective emergency communication is essential to managing hazardous materials (HazMat) incidents. Accurate reporting is the cornerstone of an organized response, enabling the swift deployment of resources and efficient onsite management. For instance, when a HazMat incident occurs, timely notification to emergency services ensures that all necessary personnel and equipment are dispatched without delay. This rapid mobilization directly affects the scene's safety and the overall efficiency of the operations.

When it comes to on-the-ground operations, clear and open channels of communication with the Incident Commander are vital. The Incident Commander plays a pivotal role in making informed decisions regarding resource allocation and risk management. Effective communication channels allow for the timely relay of hazard information, which is crucial for assessing the situation accurately. In situations where the environment may be rapidly changing due to chemical spills or other hazards, having accurate, up-to-date information helps the Incident Commander and their team make critical decisions that could save lives and protect property. Open communication ensures that any new developments at the scene, such as changes in wind direction affecting the spread of hazardous substances, are quickly transmitted and addressed.

Moreover, adhering to federal and state notification requirements is not just a legal obligation but a practical necessity for synchronizing responses across various governmental levels. These regulations have been established to ensure that local, state, and national entities can coordinate their efforts effectively during emergencies. Notifying appropriate federal agencies, for instance, facilitates access to additional resources and specialized expertise that might not be locally available. It also helps streamline communication between different jurisdictions, preventing the overlap of roles and reducing confusion. This coordination is particularly crucial in large-scale incidents that might exceed local capacity. By ensuring compliance with these notification protocols, responders help create a unified and coordinated response effort.

Guidelines for notifying emergency services typically recommend detailing the nature and scope of the incident, including specific information about the hazardous materials involved. Such details empower dispatchers and commanders to plan accordingly, deploying the right teams and resources from the start. The provision of precise information minimizes the likelihood of misinformation and its potential consequences during high-stress scenarios.

In addition to the guidelines followed by responders, communication best practices underscore the significance of having a clear message tailored to diverse audiences. During emergencies, utilizing simple and familiar language is essential to avoid misunderstandings and technical jargon that can hinder comprehension. Involved parties, such as disaster managers, must identify stakeholders like nearby communities, media outlets, and government officials, and adjust messaging to ensure it is suitable for varied groups. Clear, authoritative, and empathetic communication restores calm and directs action efficiently (Tulane University, 2023).

Ensuring that emergency communication systems are accessible to everyone in the community is a key aspect of effective communication. Disaster communication strategies should include multi-platform information dissemination, ranging from social media updates to text alerts and radio broadcasts. This diverse approach maximizes reach, catering to individuals who rely on different information channels depending on availability and trust. Integrating multiple communication avenues, such as online message boards, television, and even street signage, supports comprehensive engagement with the public. Providing regular updates through various media further establishes credibility and builds trust between responders and the community (Savoia et al., 2013).

The concept of information processing within emergency communication also emphasizes the importance of delivering content quickly yet accurately. Ensuring that information is both empirical and trustworthy prevents panic and aids in the timely adoption of protective measures by the public. Authorities must stay agile, constantly evaluating how people understand and respond to emergencies. This real-time analysis allows for the refinement of messaging and contributes to a more effective crisis response strategy (Savoia et al., 2013).

Furthermore, addressing communication inequalities is vital in reaching underserved populations who may face barriers due to language differences, cultural nuances, or disabilities. Developing messages that consider these disparities ensures broader access to life-saving information. By acknowledging these gaps, responders can adapt their communication strategies to mitigate the impact of social inequities on emergency preparedness outcomes.

Understanding and Applying the Incident Command System

The Incident Command System (ICS) plays a vital role in managing hazardous materials (HazMat) incidents by providing a standardized framework for emergency management that ensures coordinated responses across multiple agencies. Originating from efforts to address past inefficiencies and response failures, ICS is designed to be adaptable, scalable, and systematic, enabling effective incident management regardless of size or complexity.

At its core, the ICS standardizes the command structure, creating a common platform that enhances collaboration between diverse responding agencies. This standardization addresses issues of inconsistent communication and hierarchical fragmentation, which often lead to operational inefficiencies. By ensuring that all parties involved speak the same "language" and adhere to established protocols, ICS eliminates confusion and duplication of effort, thus promoting more efficient use of resources. This standardization is particularly crucial during HazMat incidents, where swift, coherent action can prevent escalation and mitigate risks to responders and the public (2023).

One of the fundamental components of ICS is its clear delineation of roles, with the Incident Commander (IC) at the helm. The IC is tasked with making critical decisions regarding resource allocation and strategic direction, thus providing focused leadership during emergencies. This role is pivotal in ensuring that all aspects of an incident are managed effectively, from logistics to operational tactics. The IC's authority to delegate tasks yet maintain overall control allows them to adapt swiftly to changing circumstances and make informed decisions based on evolving assessments of the situation (Williams et al., 2021).

Key positions within ICS also include Operations, Planning, Logistics, and Finance sections, each charged with specific responsibilities. The Operations section handles tactical fieldwork, while the Planning section focuses on developing the Incident Action Plan (IAP), assessing resources, and anticipating future needs. Meanwhile, the Logistics section provides necessary supplies and services, and the Finance section manages costs. These structured roles ensure that every aspect of the incident response is covered, reducing the likelihood of oversight or mismanagement. Within this framework, each position must communicate effectively to synchronize their efforts, thereby optimizing response time and effectiveness (2023).

Unified command is another crucial element of ICS, especially relevant in multi-jurisdictional or multi-agency incidents such as significant HazMat events. Unified command allows representatives from different agencies to work together without sacrificing individual agency authority or responsibility. This collaborative approach ensures that diverse expertise is integrated into decision-making processes, leading to more comprehensive and effective incident management. Regular briefings are a key feature of maintaining this clear communication, ensuring alignment on objectives, strategies, and ongoing coordination among responders. Such meetings provide opportunities to adjust plans based on incoming information or shifting priorities, further enhancing response efficacy (Williams et al., 2021).

Communication within ICS is not just about exchanging information but doing so using standardized procedures that everyone understands. The ICS mandates the use of plain language to avoid misunderstandings and relies on consistent terminology across all agencies. This approach minimizes errors that could arise from jargon, codes, or other ambiguous language. Proper communication ensures that all team members are aware of their responsibilities and any potential hazards present, which is particularly critical when dealing with volatile HazMat situations. By maintaining robust communication channels, ICS helps avert accidents and facilitates rapid, collective action when needed (2023).

Recognizing the importance of safety, ICS also emphasizes personal safety measures for first responders. Given the inherent dangers involved in HazMat incidents, it is crucial for responders to have clear guidelines on protective measures and procedures to follow. This includes wearing appropriate personal protective equipment (PPE) and being aware of contamination prevention practices. Ensuring responder safety through these measures not only protects lives but also maintains the integrity of the response effort by preventing incapacitation due to injury or contamination (Williams et al., 2021).

Additionally, ICS provides guidelines for post-incident analysis and learning, aiding agencies in evaluating their performance and identifying areas for improvement. This aspect of ICS ensures continuous improvement in incident management capabilities, fostering a culture of resilience and preparedness that benefits all stakeholders. Analyzing successes and shortcomings after an incident allows agencies to refine their approaches, enhance training programs, and ensure that future responses are even more effective. This cyclical process of learning and adaptation is integral to maintaining high standards of public safety and response readiness (2023).

Summary and Reflections

In this chapter, we have delved into the essential aspects of managing hazardous material incidents, emphasizing the crucial role of awareness-level responders. The chapter highlighted the significance of effective scene management, which involves a coordinated approach to control exposure and prevent incident escalation. We learned that proper use of isolation zones—Hot, Warm, and Cold—helps organize the response activities and ensures safety for both responders and the public. Tools like the Emergency Response Guidebook (ERG) aid in establishing safe distances, taking into account the type of chemicals, spill size, and weather conditions. Moreover, maintaining clear communication channels is a vital component in relaying accurate information to the Incident Commander, preventing misinformation and ensuring that strategic decisions are well-informed.

Additionally, the chapter underscored the importance of adhering to standardized protocols, such as the Incident Command System (ICS), to facilitate seamless collaboration among multiple agencies. By following these guidelines, responders can efficiently manage resources, implement containment measures, and maintain order at the scene. Training remains a cornerstone in equipping responders with the necessary skills to tackle HazMat incidents effectively. Through regular drills and exercises, first responders can refine their abilities to assess hazards, make informed decisions, and apply systematic procedures during real-world emergencies. This comprehensive understanding not only enhances preparedness but also ensures a safer outcome for every HazMat incident they encounter.

20 questions and answers based on Chapter 3: Initial Actions and Scene Management for HazMat incidents:

What is the primary role of awareness-level responders in a HazMat incident?

Answer: Awareness-level responders assess hazards, make initial decisions, and implement basic control measures to prevent escalation, ensuring safety for responders and the public.

What is the significance of effective scene management in HazMat incidents?

Answer: Effective scene management ensures safety, prevents escalation of hazards, controls exposure to harmful materials, and optimizes resource deployment.

Why is it important to control exposure to hazardous materials during a HazMat incident?

Answer: Controlling exposure helps protect responders and the public from health risks, reduces the area that needs evacuation, and minimizes environmental damage.

How do effective scene management techniques prevent incidents from escalating?

Answer: By swiftly containing the hazard, implementing control measures, and maintaining situational awareness, the risk of fires, explosions, or toxic releases is minimized.

What role do standardized protocols play in managing HazMat scenes?

Answer: Standardized protocols guide responders through each step of an incident, ensuring consistency, safety, and compliance with legal requirements.

What is the role of communication in managing a HazMat scene?

Answer: Clear and continuous communication with the incident commander and other agencies ensures coordinated efforts, quick updates, and safety measures are effectively implemented.

Why is training crucial for awareness-level responders handling HazMat incidents?

Answer: Training equips responders with the knowledge to recognize hazards, make decisions under pressure, and perform necessary actions efficiently and safely.

What are the three primary zones established during a HazMat incident?

Answer: The three primary zones are the Hot Zone (high-risk contamination area), Warm Zone (contamination reduction area), and Cold Zone (safe support area).

What is the purpose of the Hot Zone in a HazMat incident?

Answer: The Hot Zone is the area with actual or potential contamination, and it is where responders wearing proper protective gear conduct high-risk tasks like containment and rescue.

What happens in the Warm Zone during a HazMat response?

Answer: The Warm Zone serves as a buffer for decontamination processes, where responders are cleaned before moving to the Cold Zone.

What is the role of the Cold Zone in HazMat incident management?

Answer: The Cold Zone is the support area for logistics, medical treatment, and command operations, free from contamination.

How does the Emergency Response Guidebook (ERG) assist responders in HazMat incidents?

Answer: The ERG provides guidelines for initial isolation distances and protective action distances based on the type and quantity of hazardous materials involved.

What factors influence the isolation distances recommended in the ERG?

Answer: Factors include the type of hazardous material, spill size, weather conditions (e.g., wind speed), and the nature of the release.

How can weather conditions impact HazMat scene management?

Answer: Weather conditions, especially wind speed and direction, can affect how hazardous materials spread, requiring wider evacuation zones to ensure safety.

What is the importance of adapting zone boundaries during a HazMat incident?

Answer: Adapting zone boundaries based on real-time data ensures responders can stay safe as conditions change, preventing unnecessary exposure to hazards.

How does the Incident Command System (ICS) help manage HazMat incidents?

Answer: ICS provides a standardized command structure that improves communication, coordination, and decision-making, ensuring efficient management of the incident.

Who is responsible for making critical decisions during a HazMat incident under ICS?

Answer: The Incident Commander (IC) is responsible for making critical decisions regarding strategy, resource allocation, and operational direction.

What is Unified Command, and why is it important in HazMat incidents?

Answer: Unified Command allows multiple agencies to collaborate effectively, ensuring that all aspects of the incident are managed without conflicting priorities or confusion.

How does the ICS structure prevent confusion during a HazMat response?

Answer: The ICS uses clear roles and standardized communication, ensuring that everyone involved understands their responsibilities and tasks, reducing errors.

Why is post-incident analysis a key element of ICS?

Answer: Post-incident analysis helps identify successes and areas for improvement, ensuring that future responses are more efficient and better coordinated.

Chapter 4: Defensive Actions and Protective Measures

(Operations Level)

Defensive actions and protective measures are essential components in managing hazardous materials (HazMat) incidents. Understanding these concepts is crucial for those involved in hazardous response operations, such as firefighters, first responders, and hazardous materials students. This chapter delves into the strategic importance of implementing defensive actions to mitigate potential risks. It emphasizes the need for safety protocols that protect both the individuals directly involved in the response and the broader community. By focusing on the tactical elements required to handle HazMat situations effectively, this chapter provides a comprehensive overview of how structured defensive responses can prevent incidents from escalating.

Throughout the chapter, readers will explore various strategies and techniques employed during HazMat incidents. It covers containment methods such as absorption, damming, and diverting, which aim to control the spread of hazardous substances. Additionally, there is an examination of the coordination between responders and the Incident Command System (ICS) to ensure a unified and efficient approach. The chapter also discusses the roles and responsibilities of operations-level responders, emphasizing their training and equipment limitations. Furthermore, the importance of proper communication and maintaining clear channels with the public during emergencies is highlighted. By the end of the chapter, readers will have gained valuable insights into the systematic approaches required to manage HazMat incidents safely and effectively.

Introduction to Defensive Actions and Protective Measures

The importance of defensive actions within hazardous materials (HazMat) response operations cannot be overstated. These actions play a critical role in mitigating risks associated with HazMat incidents, ensuring safety, and preserving life, property, and the environment. Defensive actions are essential because they minimize the spread of hazardous materials, protect responders and the public, and operate within established training and personal protective equipment (PPE) limits while coordinating effectively with the Incident Command System (ICS).

One of the primary objectives of defensive actions is to minimize the spread of hazardous materials. When a HazMat incident occurs, the potential for these materials to disperse into the surrounding environment poses significant challenges. Initial responders at the operational level are trained to take actions that contain a release from a safe distance and prevent its spread. According to the HAZWOPER standard, first responders are not required to stop the release but to focus on controlling it defensively (<i>IN CASE of HAZARDOUS SPILL: FIRST RESPONDER ROLE & INITIAL RESPONSE ACTIONS</i>, 2025). This strategy encompasses techniques like absorption, damming, diverting, vapor dispersion, and containment. By employing these methods, responders can restrict the movement of hazardous substances and significantly reduce the risk of exposure to nearby populations.

Strategic actions taken during defensive operations also aim to protect responders, the public, and the environment from the detrimental effects of hazardous spills. Responders prioritize safeguarding human health and preserving natural resources through coordinated efforts. The deployment of effective defensive measures ensures that incidents are controlled with minimal harm. For instance, evacuees might be directed away from potentially contaminated areas as part of an evacuation plan, allowing for safer zones around the release site (SeventhQueen, 2021). Maintaining clear communication channels between responders and the public is vital, ensuring everyone is informed and aware of any evolving hazards. Using well-established communication protocols aids in promoting trust and cooperation among stakeholders.

Another crucial aspect of defensive HazMat operations is adherence to operational training and PPE limits. Operations-level personnel are equipped with specific skills and gear appropriate to their role, enabling them to execute actions safely without compromising their own well-being. They are trained to understand the limitations of their equipment and operate within those boundaries, avoiding exposure to high-risk zones or situations where their PPE might be inadequate (<i>IN CASE of HAZARDOUS SPILL: FIRST RESPONDER ROLE & INITIAL RESPONSE ACTIONS</i>, 2025). Effective training equips them with knowledge concerning the use of PPE, such as respiratory protection and chemical suits, ensuring that responders remain shielded during operations.

Coordination with the ICS is an indispensable component of successful defensive operations in HazMat incidents. The ICS provides a structured framework for managing emergency responses by defining roles and responsibilities, thereby facilitating seamless collaboration among responders from different agencies. Operations-level responders work closely with the ICS to ensure that all defensive actions align with broader strategic goals. This includes participating in joint planning meetings, sharing information, and providing updates on the progress of containment efforts. Such coordination is crucial in optimizing resource allocation, enhancing situational awareness, and ensuring that defensive actions achieve desired outcomes (SeventhQueen, 2021).

Role of Operations-Level Responders

In the realm of hazardous materials (HazMat) incidents, operations-level responders play a pivotal role in ensuring safety and mitigating risks. These responders are tasked with implementing defensive actions rather than engaging in direct mitigation efforts, making their responsibilities distinct yet crucial within incident management. The foundation of their duties lies in recognizing and assessing HazMat threats accurately and efficiently.

Understanding the Operations-Level Role in HazMat Incidents involves identifying potential hazards through the use of tools like placards and Safety Data Sheets (SDS). Placards serve as external identifiers on transport vehicles, providing vital information about the contents being carried. Each placard is color-coded and displays symbols that denote the class of hazard, such as flammable, corrosive, or toxic substances. Proper interpretation of these symbols allows responders to make informed decisions quickly. SDS, on the other hand, provide detailed information about chemical properties, health hazards, storage guidelines, and emergency measures. Responders must be adept at extracting pertinent data from SDS to guide their initial actions and ensure their safety as well as that of others involved in the incident (<i>1926.65 - Hazardous Waste Operations and Emergency Response. | Occupational Safety and Health Administration</i>, 2019).

Moreover, utilizing the Emergency Response Guidebook (ERG) becomes essential for setting response guidelines. The ERG offers a structured approach to handling various hazardous situations by outlining potential dangers and recommended protective measures. By referencing the ERG swiftly, responders can determine safe distances, isolation techniques, and specific actions tailored to particular hazards, thereby enhancing their ability to manage the situation effectively.

Another integral responsibility of operations-level responders is supporting technician-level personnel through isolation and evacuation efforts. When a HazMat incident presents an immediate threat to public health or property, prompt evacuation is necessary. Operations-level responders assist by establishing perimeters and guiding affected individuals to safety. They also coordinate with incident command structures to ensure evacuations are orderly and efficient, minimizing panic and maintaining control over the situation.

Isolation, another critical component, involves creating barriers to prevent the spread of hazardous substances. This may include sealing off areas or using containment devices to restrict movement and exposure (<i>California Code of Regulations, Title 8, Section 5192. Hazardous Waste Operations and Emergency Response.</i>, 2014).

Beyond immediate containment and evacuation, a fundamental objective is implementing strategies to protect life, property, and the environment. Operations-level responders must assess the potential impact of hazardous materials on surroundings and devise plans to minimize damage. This could entail deploying absorbents to contain spills, using water sprays or foam to mitigate vapors, or deploying ventilation systems to disperse harmful gases. Such strategies require not only technical knowledge but also a clear understanding of environmental factors and the potential consequences of different actions.

In addition to reactive measures, proactive strategies form a core aspect of an operations-level responder's duties. This involves ongoing collaboration and communication with other emergency response teams and agencies. By maintaining liaison with local authorities, fire departments, and hazardous materials experts, responders can access additional resources and expertise, enhancing their capacity to manage complex scenarios. Through joint training exercises and tabletop simulations, operations-level responders refine their skills and preparedness, ensuring their readiness when real incidents occur.

A key element in the effectiveness of operations-level responders is adherence to safety protocols and proper use of Personal Protective Equipment (PPE). Ensuring that all team members are equipped with appropriate PPE minimizes the risk of exposure to hazardous substances. Regular training on donning and doffing procedures, equipment maintenance, and decontamination processes is imperative to maintaining high safety standards.

Furthermore, decision-making in dynamic environments relies on information accuracy, which underscores the importance of continuous monitoring and assessment. Utilizing advanced detection equipment, such as portable gas detectors or chemical analyzers, enables responders to gather real-time data on the nature and concentration of hazardous substances present. This information is crucial for adjusting response strategies as conditions evolve.

Recognizing that HazMat incidents often unfold unpredictably, operations-level responders must be adaptable and resourceful. In some cases, unforeseen complications or secondary hazards might arise, necessitating quick thinking and problem-solving abilities. Responders must prioritize tasks, assess risks continually, and make swift adjustments to safeguard lives and property effectively.

Ultimately, the responsibility of operations-level responders extends beyond immediate crisis management to foster community resilience. Engaging with community leaders and participating in public outreach initiatives helps raise awareness about the risks associated with hazardous materials and the importance of safety precautions. By educating the public on emergency procedures and encouraging preparedness, responders contribute to building a safer and more informed society.

Defensive Strategies: Containment, Confinement, Exposure Reduction

In responding to hazardous materials (HazMat) incidents, first responders, firefighters, and hazardous materials students must be equipped with a clear understanding of defensive strategies. These strategies are essential in managing the potentially devastating effects of hazardous substances on people, property, and the environment.

The primary goal of defensive actions in HazMat operations is to contain and control the spread of hazardous substances effectively while ensuring maximum safety for both responders and the public. One key strategy employed is containment, which involves using barriers to restrict the movement of spilled materials. Techniques such as diking and damming are vital tools in this endeavor. Diking involves creating barriers around the spill using soil or other materials to prevent further spreading. Similarly, damming is used to direct or block the flow of liquid hazards, providing a crucial means to manage spills efficiently. Implementing these methods helps limit environmental contamination and reduces the potential for more extensive damage. It is also critical to continually assess the effectiveness of these barriers throughout the incident response and make necessary adjustments as situations evolve.

Confinement techniques are another pivotal component of defensive actions in HazMat scenarios. These measures focus on controlling the vapor or gases released from hazardous materials. Vapor suppression is one such technique commonly used to minimize exposure risks. This involves applying foam or other agents onto liquid surfaces to reduce vapor emissions, preventing them from becoming airborne hazards. Dispersion, on the other hand, involves diluting harmful vapors into the atmosphere to lower their concentration and potential impact. This may include using water sprays or fans to disperse clouds of toxic vapors effectively. The choice between vapor suppression and dispersion depends on factors such as wind direction, weather conditions, and the nature of the hazardous material involved.

Furthermore, exposure reduction measures are integral to defensive strategies. The principles of time, distance, and shielding form the basis of these measures, offering critical guidance in minimizing responders' exposure to hazardous environments. Limiting the amount of time spent in the vicinity of a hazard significantly decreases the risk of adverse health effects. Responders should aim to complete tasks swiftly and efficiently, while rotating personnel regularly to prevent prolonged exposure. Distance plays an equally pivotal role: maintaining a safe distance from the source of contamination ensures reduced exposure levels. Establishing exclusion zones and cording affected areas are effective ways to enforce safe distances during an incident. Additionally, the use of personal protective equipment (PPE) and positioning behind physical barriers contribute to effective shielding from hazards.

Beyond these tactical approaches, it is crucial to emphasize the importance of thorough training and adherence to established guidelines. Responders must be familiar with standard operating procedures and consistently apply them during incidents to ensure effectiveness and safety. Understanding the characteristics of different hazardous materials, their behaviors, and potential health impacts is fundamental for informed decision-making in the field. First responders should also engage in continuous education to stay updated on best practices and emerging technologies that can enhance defensive capabilities.

Supportive evidence for these strategies can be found in existing safety standards and regulations. For instance, the Occupational Safety and Health Administration (OSHA) highlights the significance of respiratory protection, personal protective equipment, and initial medical surveillance in mitigating exposure risks (<i>OSHA Technical Manual (OTM) - Section v: Chapter 3 | Occupational Safety and Health Administration</i>, 2025). Furthermore, the use of engineering controls, such as ventilation systems and sealed containment structures, underscores the preference for mechanical means in reducing airborne hazards. Similarly, guidelines for handling incidental hazardous substance releases are delineated by federal regulations (<i>Traffic Incident Management in Hazardous Materials

Spills in Incident Clearance: 4.0 Hazard Materials Incident Clearance Compliance Requirements - FHWA Office of Operations</i>, 2025), stressing the need for meticulous adherence to safety procedures to minimize health and safety threats.

Personal Protective Equipment (PPE) and Decontamination Principles

Understanding the correct selection and application of personal protective equipment (PPE) and decontamination practices is crucial for safety in hazardous materials (HazMat) incidents. First responders, firefighters, and HazMat students must comprehend the need to choose adequate PPE levels according to specific hazards they might encounter. As no single PPE suits all hazards, PPE serves as one piece of a multifaceted approach toward safety that includes exposure control procedures and equipment (Personal Protective Equipment (PPE) - CHEMM, 2021).

Selection begins with understanding different PPE levels. Ensuring appropriate protection against chemical, radiological, or biological hazards requires careful assessment. Level A protection, involving fully encapsulating chemical suits and self-contained breathing apparatus, offers the highest level of safety when confronting extreme risks such as high concentrations of toxic substances. On the other hand, Level D offers minimal protection, suitable for nuisance contamination only (Personal Protective Equipment (PPE) - CHEMM, 2021). The incident commander or safety officer typically determines PPE requirements based on situational hazard assessments.

Following strict donning and doffing procedures is equally significant in preventing contamination. Donning involves correctly wearing PPE to ensure intended protection, while doffing means safely removing it to avoid self-contamination. Improper handling during these processes can lead to hazardous exposure (CDC, 2023). For example, not following donning guidelines like checking the integrity of suits before wearing or failing to adhere to protocols in contaminated zones can compromise safety.

In hazardous scenarios, maintaining clear zone demarcations is essential. The exclusion zone, also known as the hot zone, poses the highest risk due to potential contamination. Here, proper PPE is mandatory. Transitioning into the warm zone allows for initiating decontamination activities, helping reduce contamination between high-risk and safe areas. Finally, the cold zone remains contamination-free, serving as a planning and recovery space (CDC, 2023).

Decontamination methods are crucial in minimizing secondary exposure risks. Executing these processes effectively ensures hazardous substances do not transfer from one person or object to another. Decontamination may involve simple rinsing procedures or more extensive chemical neutralization, depending on the contaminant type and exposure severity. Implementing effective decontamination practices is vital for individual safety and environmental protection, reducing the likelihood of hazardous material spreading beyond the incident site.

Confinement strategies play an instrumental role in controlling the spread of contaminants. By implementing confinement techniques such as establishing containment perimeters or using absorbent materials to contain spills, responders can limit exposure risks until proper decontamination is conducted.

Comprehensive PPE use relies on having a structured management program, ensuring certifications like respirator fit testing, and understanding PPE limitations, such as restricted mobility and communication constraints. Realizing these limitations helps define operational boundaries and safety measures required when operating under restrictive PPE conditions. Additionally, understanding psychological factors, like stress induced by confined space within PPE, is critical for anticipating and managing responder health impacts (Personal Protective Equipment (PPE) - CHEMM, 2021).

For first responders and emergency personnel, ongoing training in PPE use and decontamination practices maintains skill proficiency needed for responding effectively in HazMat incidents. Understanding each PPE element's role – from respirators to eye protection – forms the foundation of this training. Regular drills and educational updates ensure responders are familiar with new equipment standards and decontamination advancements.

Furthermore, responders are encouraged to adopt prudent exposure reduction measures, including limiting time in contaminated zones, maximizing distance from hazards, and using physical shields when practical. These exposure reduction measures serve as fundamental practices alongside PPE to enhance overall safety and effectiveness in hazardous environments (Personal Protective Equipment (PPE) - CHEMM, 2021).

Summary and Reflections

This chapter has explored the vital role of defensive actions in hazardous materials (HazMat) incidents, emphasizing the importance of minimizing risks to individuals, communities, and the environment. We have delineated the strategies employed by operations-level responders to contain and control hazardous substances effectively, ensuring safety throughout the incident. By focusing on techniques such as containment, confinement, and exposure reduction, responders are equipped with the knowledge and tools needed to address potential threats while safeguarding themselves and those around them. This structured approach is reinforced through thorough training, familiarity with Personal Protective Equipment (PPE), and adherence to established guidelines, which collectively enhance the efficacy of HazMat response efforts.

Understanding the responsibilities and limitations of operations-level responders emphasizes the need for clear communication and coordination with other emergency roles. The integration of established frameworks like the Incident Command System (ICS) ensures that all actions align with broader strategic objectives, optimizing resource allocation and enhancing situational awareness. As we conclude this chapter, it's crucial for first responders, firefighters, and HazMat students to recognize and implement these defensive measures consistently. By doing so, they contribute to a safer environment for all, responding adeptly to the dynamic challenges that HazMat incidents often present, and ultimately fostering greater resilience within their communities.

20 questions and answers based on Chapter 4: Defensive Actions and Protective Measures from the provided text:

What is the primary objective of defensive actions in HazMat incidents?

Answer: The primary objective is to minimize the spread of hazardous materials and protect responders, the public, and the environment.

What techniques are used to control the spread of hazardous materials?

Answer: Techniques include absorption, damming, diverting, vapor dispersion, and containment.

What role do operations-level responders play during a HazMat incident?

Answer: Operations-level responders implement defensive actions to contain and control the spread of hazardous materials, ensuring safety without engaging in direct mitigation.

What does the HAZWOPER standard require first responders to do in a HazMat incident?

Answer: First responders are required to control the release of hazardous materials defensively, not to stop the release.

What is the role of the Incident Command System (ICS) in HazMat response?

Answer: ICS provides a structured framework for managing the emergency response, ensuring coordination and alignment of defensive actions with broader strategic goals.

What are some common containment methods used in HazMat incidents?

Answer: Diking and damming are common methods used to prevent the spread of hazardous materials.

What does vapor suppression involve in HazMat response?

Answer: Vapor suppression involves applying foam or other agents to liquid surfaces to reduce the emission of harmful vapors into the atmosphere.

How does vapor dispersion differ from vapor suppression?

Answer: Vapor dispersion involves diluting harmful vapors into the atmosphere to lower their concentration, while vapor suppression reduces vapor emissions directly from the source.

Why is the use of Personal Protective Equipment (PPE) critical during HazMat incidents?

Answer: PPE protects responders from exposure to hazardous substances, ensuring their safety during operations.

What is the purpose of the Emergency Response Guidebook (ERG) during HazMat response?

Answer: The ERG provides responders with guidelines on how to handle hazardous materials, including safety distances, isolation techniques, and specific actions for various hazards.

How do operations-level responders assist with evacuation efforts?

Answer: They establish perimeters, guide affected individuals to safety, and coordinate with incident command to ensure orderly and efficient evacuations.

What are the primary goals of operations-level responders during a HazMat incident?

Answer: The primary goals are to minimize damage, protect life, property, and the environment, and to implement strategies like containment and evacuation.

How can exposure to hazardous materials be minimized during a HazMat incident?

Answer: Exposure can be minimized by limiting the time spent in hazardous zones, maintaining safe distances, and using appropriate PPE.

What is the significance of proper communication during a HazMat incident?

Answer: Clear communication ensures that all stakeholders, including responders and the public, are informed about evolving hazards, which promote trust and cooperation.

What are the three primary principles of exposure reduction in HazMat operations?

Answer: The principles are time, distance, and shielding, which are used to minimize exposure to hazardous materials.

Why is continuous monitoring crucial during a HazMat incident?

Answer: Continuous monitoring provides real-time data on the nature and concentration of hazardous substances, allowing responders to adjust strategies accordingly.

What is the purpose of decontamination during a HazMat incident?

Answer: Decontamination helps reduce the transfer of hazardous materials from one person or object to another, minimizing the risk of secondary exposure.

What are Level A PPE requirements for HazMat responders?

Answer: Level A PPE involves fully encapsulating chemical suits and self-contained breathing apparatus, providing the highest level of protection for responders.

What is the difference between the hot, warm, and cold zones in HazMat incidents?

Answer: The hot zone is the highest risk area requiring full PPE, the warm zone is where decontamination occurs, and the cold zone is contamination-free and used for planning and recovery.

What is the significance of joint training exercises for operations-level responders?

Answer: Joint training exercises help responders refine their skills, improve coordination with other agencies, and prepare for real HazMat incidents.

Chapter 5: Chemical and Physical Properties of Hazardous Materials

(Operations Level)

Understanding the chemical and physical properties of hazardous materials is essential for ensuring safety during emergencies. In this chapter, we delve into the diverse reactions and impacts these materials exhibit under varying conditions. By recognizing their complex behaviors, responders can adopt effective strategies to manage incidents safely and efficiently. The dynamic nature of hazardous substances requires a thorough comprehension of how they interact with environmental factors, thereby dictating the appropriate response measures.

This chapter will explore several key topics related to hazardous material management at an operations level. We will examine how different properties, such as flammability, toxicity, and reactivity, influence emergency response strategies. Additionally, the chapter discusses routes of exposure, highlighting how hazardous materials pose risks to responder and public health. Finally, we will address the environmental consequences of mishandling these substances and outline detection methods crucial for monitoring and mitigation efforts. Through detailed discussions, the chapter aims to equip students, firefighters, and first responders with the knowledge necessary to respond effectively in emergency situations involving hazardous materials.

Importance of Understanding HazMat Behavior

Comprehending the behavior of hazardous materials is paramount for those involved in emergency response, including students and professionals managing such incidents. Diverse materials exhibit distinct reactions under different conditions, directly affecting how incidents are managed. Recognizing these behaviors offers a foundation for immediate and effective action during emergencies.

Different materials react uniquely depending on their properties and environmental conditions. For instance, temperature changes can lead to significant alterations in a material's state or reactivity. A chemical that remains stable at room temperature might become highly reactive when exposed to higher temperatures or moisture. This variability necessitates a nuanced understanding among responders to anticipate possible scenarios and prepare accordingly. Awareness of potential outcomes helps in developing robust strategies, ensuring responders address each situation with precision and foresight.

The dynamic interplay between physical and chemical properties of hazardous substances determines the appropriate strategy for containment and responder safety. Physical properties such as boiling point, vapor pressure, and density influence how a substance may behave in an incident. Chemical properties, such as acidity, basicity, and oxidation potential, dictate how it reacts with other materials and its environment. For example, a high vapor pressure indicates a gas will likely disperse quickly in the air, potentially requiring immediate evacuation measures to ensure public and responder safety.

These properties also play a vital role in establishing personal protective equipment (PPE) requirements for responders. Understanding whether a substance is corrosive or flammable guides the selection of suitable protective gear, minimizing risks of injury or illness among emergency personnel. Responders equipped with knowledge of these properties can tailor their approach to containment, choosing methods that prevent further hazards—like using non-sparking tools around flammable vapors to avoid ignition.

Moreover, comprehensive knowledge of hazardous materials significantly influences mitigation measures and communication during incidents. When a substance's properties are well-understood, responders can implement specific mitigation techniques promptly. For instance, knowing that certain chemicals react violently with water allows responders to avoid using water-based suppression systems, opting instead for dry chemical extinguishers or other suitable alternatives. This informed decision-making extends to communicating effectively with team members and the public about potential dangers and necessary precautions, ensuring coordinated and efficient responses.

Effective communication within teams and with the public is crucial during emergencies involving hazardous materials. Clear communication, rooted in a thorough understanding of hazardous materials' properties, aids in disseminating accurate information efficiently. By providing appropriate warnings and instructions, responders minimize panic and facilitate orderly evacuations or shelter-in-place orders, ultimately enhancing overall safety.

Preventing exposure to hazardous materials not only safeguards responders but ensures public safety in emergencies. Preparation based on an understanding of these materials' behaviors leads to quick and decisive action, aimed at containing leaks, spills, or releases before they escalate. Implementing preventive measures like barriers or ventilation systems can avert widespread exposure, protecting communities and ecosystems from harm.

Such prevention aligns with regulatory standards like OSHA's HAZWOPER (Hazardous Waste Operations and Emergency Response), which mandates stringent procedures to protect workers dealing with hazardous waste operations and emergency responses. Compliance with these standards requires a deep understanding of hazardous materials, underscoring their role in shaping effective preparedness and response strategies. Responders trained under

HAZWOPER are equipped with knowledge and skills essential for addressing hazardous situations safely and efficiently (<i>Hazardous Waste Operations and Emergency Response (HAZWOPER) - Overview | Occupational Safety and Health Administration</i>, 2025).

Flammability, Toxicity, and Reactivity

The behavioral characteristics of hazardous materials are pivotal in determining effective emergency response and ensuring the safety of responders and the public. One fundamental aspect to consider is flammability and combustibility, which can significantly impact fire risks and guide response strategies. These properties are often measured by flash points—the lowest temperature at which a liquid gives off sufficient vapor to ignite in air. For instance, a material with a low flash point is considered highly flammable and poses an immediate fire hazard. Understanding this property helps in assessing the appropriate firefighting techniques and protective measures necessary to manage such risks effectively. Evaluating flash points allows responders to determine safe handling procedures and storage requirements, minimizing potential ignition sources and guiding the choice of extinguishing agents.

Moving on to toxicity, it presents both acute and chronic health risks that dictate exposure limits and influence incident planning. Acute toxicity refers to short-term effects following exposure, while chronic toxicity relates to long-term consequences from repeated exposure over time. These health hazards are categorized based on permissible exposure limits established to protect workers and the public. Such limits play a critical role in forming safety protocols, including the use of personal protective equipment (PPE) and decontamination procedures. Effective incident planning should involve comprehensive risk assessments to evaluate potential toxic exposures and develop mitigation strategies. This includes understanding systemic effects like carcinogenicity or neurotoxicity, and target organ effects on the liver, kidneys, or respiratory system, as outlined by regulatory guidelines (<i>Hazard Communication - Guidance for Hazard Determination | Occupational Safety and Health Administration</i>, 2023).

Chemical reactivity is another vital factor, as it can lead to dangerous reactions if not properly managed. Understanding endothermic and exothermic processes is essential for anticipating and mitigating these risks. Exothermic reactions release heat, potentially causing explosions or fires, while endothermic reactions absorb heat, possibly resulting in cooling and condensation issues. The awareness of chemical stability and reaction thresholds ensures that first responders and handlers recognize signs of instability in materials. For example, storing incompatible materials together can trigger unwanted exothermic reactions, illustrating why segregating reactive chemicals is crucial. Comprehensive training and preparation involve knowing how to handle unstable substances safely, utilizing specialized equipment, and enacting appropriate emergency procedures when necessary.

Identifying incompatible materials helps avoid hazardous mixing scenarios and prevent potential catastrophic events. Proper segregation and storage of chemicals are vital to prevent them from reacting adversely. Common groups that should not be stored together include acids with bases, as mixing them can produce violent exothermic reactions and release harmful gases. Similarly, oxidizers should be kept away from flammable materials to prevent accidental ignition or combustion. An effective chemical storage method involves categorizing chemicals based on their reactive properties and consulting safety data sheets (SDS) for proper storage guidelines (<i>Chemical Storage: Categories, Hazards and Compatibilities | Protocol</i>, n.d.). This practice is crucial during emergencies, such as fires or spills, where the unintentional mixing of chemicals could result in toxic fume production or explosions. Training personnel to recognize incompatibility and adhering to storage protocols significantly reduces the likelihood of such incidents.

Routes of Exposure and Health Effects

Hazardous materials pose significant health risks, primarily through various routes of entry into the body. Understanding these pathways is crucial for first responders, firefighters, and hazardous materials students to ensure effective safety practices and timely interventions. Inhalation represents the most common route of exposure, posing severe respiratory threats. When inhaled, chemicals enter the respiratory tract and can be absorbed into the bloodstream, leading to widespread systemic effects. Symptoms might include irritation in the eyes, nose, and throat, coughing, difficulty breathing, dizziness, and even collapse in extreme cases. Therefore, ensuring proper ventilation in hazardous environments and using appropriate respiratory protective equipment is essential for safeguarding against inhalation risks. Utilization of fume hoods, dust masks, and respirators can effectively mitigate these dangers (<i>7.4 Routes of Chemical Entry</i>, n.d.; <i>7.4 Routes of Chemical Entry | Environment, Health and Safety</i>, n.d.).

Additionally, skin absorption is a critical concern when dealing with hazardous materials. Chemicals can penetrate the skin barrier, causing local or systemic toxicity. This route necessitates the consistent use of Personal Protective Equipment (PPE), such as gloves, goggles, and protective clothing, to prevent direct contact with harmful substances. It is vital to understand that the effectiveness of PPE depends on its proper use and maintenance, making regular checks and adherence to guidelines indispensable for reducing exposure risks.

While ingestion may occur less frequently compared to inhalation and skin absorption, it still presents notable hazards. Contaminants can transfer from hands or surfaces to the mouth, especially in environments where food or beverages are consumed near chemical handling areas. Practicing good personal hygiene, such as thorough hand washing before eating or drinking, significantly reduces the likelihood of accidental ingestion. Additionally, clear labeling and segregating food storage from chemicals can further minimize inadvertent ingestion incidents. The effects of ingestion might not be immediately apparent, but long-term exposure can lead to chronic health issues, emphasizing the need for diligent precautionary measures (<i>7.4 Routes of Chemical Entry</i>, n.d.; <i>7.4 Routes of Chemical Entry | Environment, Health and Safety</i>, n.d.).

Injection, though relatively rare, poses severe consequences due to the potential for direct introduction of chemicals or pathogens into the bloodstream. This can occur through punctures or cuts from contaminated sharp objects like glass, needles, or blades. Immediate symptoms may develop, making quick response and medical intervention a priority. Wearing protective gloves and handling sharp objects with care are primary strategies to prevent injection incidents. Moreover, properly disposing of sharp items in designated containers helps protect all personnel within the environment (<i>7.4 Routes of Chemical Entry</i>, n.d.; <i>7.4 Routes of Chemical Entry | Environment, Health and Safety</i>, n.d.).

Understanding these routes of exposure enables individuals working with hazardous materials to implement effective prevention strategies. Beyond individual actions, institutional policies and training programs play a pivotal role in minimizing exposure risks. Establishing clear protocols for emergency situations, regular training sessions, and ensuring compliance with safety standards form the foundation of a safe working environment. Through education and adherence to safety practices, the health impacts associated with hazardous materials can be significantly reduced, protecting both individuals and communities involved in incident management.

Environmental Impacts and Detection Methods

Understanding the environmental consequences of hazardous materials and the methods to monitor them is crucial for effective management and response. Spills, leaks, and improper handling can lead to soil and water contamination, which poses significant risks to ecosystems and human health. When hazardous substances enter the environment, whether through industrial activities, mishandling, or accidental releases, they can result in far-reaching impacts.

Soil contamination occurs when hazardous substances infiltrate the soil, altering its quality and functionality. Contaminated soil can disrupt agricultural productivity and hinder the growth of plants, making it difficult to cultivate crops. The presence of harmful chemicals like heavy metals, pesticides, and petrochemicals can affect plant metabolism and potentially enter the food chain, posing health risks to animals and humans who rely on these resources. Moreover, contaminated soil can leach toxins into groundwater, threatening drinking water sources and further degrade the ecological health of surrounding areas (Team, 2024).

Water pollution represents another dire consequence of hazardous material mismanagement. Toxic chemicals from spills can infiltrate rivers, lakes, and other bodies of water. This disruption not only affects aquatic life by leading to a loss of biodiversity but also presents direct threats to human communities relying on these waters for daily use. Harmful substances can accumulate in fish and plants, creating bioaccumulation issues that can make food products unsafe for consumption. The ripple effect through ecosystems underscores the need for robust management practices (Team, 2024).

Air pollution resulting from toxic vapors emitted by hazardous materials can have long-lasting repercussions on atmospheric conditions. Volatile organic compounds (VOCs) and other pollutants contribute to the formation of smog and degrade air quality. These emissions can exacerbate respiratory issues in vulnerable populations and contribute to broader public health concerns. Understanding the source and spread of these pollutants requires careful analysis and continuous monitoring, especially in urban areas where air quality is already a pressing concern (Abubakar et al., 2022).

To effectively mitigate these environmental risks, proper air monitoring and detection methods are necessary. Diverse equipment, such as gas detectors and air sampling devices, play a critical role in ensuring real-time safety assessments. Accurate monitoring enables emergency responders to identify the presence of hazardous gases promptly, assess their concentration levels, and take appropriate measures to protect public health. Training personnel in the operation and interpretation of these instruments is essential for maintaining readiness in the face of potential emergencies.

Proper guidelines for air monitoring and detection methods include selecting the right equipment based on the types of hazardous materials present, understanding the proper calibration of sensors, and knowing how to interpret data results. Regular updates to technology and methodology enhance preparedness and response capabilities, providing an added layer of protection against unforeseen incidents.

Moreover, compliance with waste disposal regulations and remediation strategies is vital for site recovery and sustainability. Strict adherence to guidelines ensures that hazardous materials are disposed of in a manner that does not jeopardize public health or the environment. Remediation efforts, such as soil washing, containment, and bioremediation, aim to restore contaminated sites to a condition suitable for future use. These efforts are not only necessary for environmental restoration but also for economic viability, allowing lands to be reused for agriculture, recreation, or development (Abubakar et al., 2022).

Bringing It All Together

Understanding the behaviors of hazardous materials is crucial for developing effective emergency response strategies and ensuring safety. This chapter focused on how the unique physical and chemical properties of these materials influence their reactivity, flammability, and toxicity under various conditions. Recognizing how factors such as temperature and environmental changes affect these materials allows responders to predict potential scenarios and prepare accordingly. By grasping these concepts, students, firefighters, and first responders are better equipped to manage incidents with precision and foresight, aligning their actions with established safety standards.

The detailed discussion also emphasized the importance of personal protective equipment (PPE) and communication during emergencies. Knowing which PPE to use according to a material's properties ensures that responders minimize exposure risks and protect themselves effectively. Additionally, clear and accurate communication based on an understanding of hazardous materials' behavior aids in coordinating team actions and informing the public about necessary precautions. Such comprehensive knowledge supports quick, decisive action aimed at containing incidents, preventing widespread exposure, and protecting communities and ecosystems from harm.

20 questions based on Chapter 5: Chemical and Physical Properties of Hazardous Materials (Operations Level):

What is the importance of understanding the chemical and physical properties of hazardous materials during an emergency?

Answer: It helps responders predict material behavior under various conditions, enabling them to implement effective strategies for containment, safety, and communication during hazardous material incidents.

How can temperature affect the properties of hazardous materials?

Answer: Temperature changes can alter a material's state, reactivity, or volatility. For example, a chemical that remains stable at room temperature may become highly reactive or volatile when exposed to higher temperatures.

What role do physical properties like boiling point, vapor pressure, and density play in emergency response?

Answer: These properties influence how a substance behaves during an incident, affecting decisions on containment methods and responder safety, such as choosing appropriate extinguishing agents or PPE.

Why is knowledge of chemical properties such as acidity and oxidation potential crucial for responders?

Answer: Chemical properties dictate how substances interact with each other and their environment, helping responders anticipate dangerous reactions and select appropriate response methods.

How does a high vapor pressure affect the response to a hazardous material incident?

Answer: A high vapor pressure indicates a substance will disperse quickly as a gas, potentially creating a fire or explosion hazard, requiring immediate evacuation and specialized response techniques.

What is the significance of flash points in relation to hazardous materials?

Answer: The flash point is the lowest temperature at which a liquid gives off vapor to ignite in air. Substances with low flash points are highly flammable and present immediate fire risks, guiding firefighting strategies.

What is the difference between acute and chronic toxicity in hazardous materials?

Answer: Acute toxicity refers to short-term health effects from immediate exposure, while chronic toxicity refers to long-term health issues from repeated exposure over time.

What is the role of permissible exposure limits (PELs) in hazardous materials management?

Answer: PELs are regulatory limits on the amount or concentration of a hazardous substance in the air, helping protect the health of workers and the public by setting safe exposure thresholds.

What are exothermic and endothermic reactions, and why are they important for responders to understand?

Answer: Exothermic reactions release heat, potentially causing fires or explosions, while endothermic reactions absorb heat. Knowing these types of reactions helps responders anticipate risks and manage them accordingly.

Why is it important to segregate incompatible chemicals during storage and emergencies?

Answer: Incompatible chemicals can react violently, causing fires, explosions, or toxic gas release. Proper segregation prevents these dangerous reactions and enhances safety.

What is the most common route of exposure to hazardous materials, and what protective measures should be used?

Answer: Inhalation is the most common route of exposure. To protect against inhalation risks, responders should use respirators, fume hoods, or other respiratory protective equipment.

How does skin absorption pose a threat in hazardous material incidents?

Answer: Chemicals can penetrate the skin and cause local or systemic toxicity. PPE like gloves, goggles, and protective clothing is crucial to prevent skin exposure.

Why is ingestion of hazardous materials a concern, and how can it be prevented?

Answer: Ingestion can lead to long-term health issues. Preventing ingestion involves good personal hygiene, such as thorough handwashing, and clear segregation of food and chemical handling areas.

What are the risks associated with the injection route of exposure to hazardous materials?

Answer: Injection can introduce chemicals or pathogens directly into the bloodstream, potentially causing immediate and severe health effects. Protective gloves and proper handling of sharp objects help reduce the risk.

What are the environmental consequences of hazardous material spills or leaks?

Answer: Spills or leaks can contaminate soil, water, and air, affecting ecosystems, wildlife, and human health. They can disrupt agriculture, pollute water sources, and lead to air quality issues.

How can hazardous material spills contaminate soil, and what are the consequences?

Answer: Hazardous materials can seep into the soil, affecting plant growth and entering the food chain. Contaminated soil may also leach toxins into groundwater, threatening drinking water sources.

What is bioaccumulation, and how does it relate to hazardous material spills?

Answer: Bioaccumulation occurs when toxic substances accumulate in the tissues of organisms, leading to health risks for animals and humans who consume contaminated food or water.

Why is air monitoring crucial during a hazardous material incident?

Answer: Air monitoring helps detect hazardous gases and assess their concentration levels, enabling responders to take protective measures and mitigate health risks to the public.

What are some key methods for detecting hazardous materials in the air?

Answer: Gas detectors, air sampling devices, and continuous monitoring systems are essential tools for detecting hazardous gases and evaluating their concentrations in real-time.

What is the role of remediation in managing hazardous material incidents, and what strategies are used?

Answer: Remediation involves cleaning up contaminated sites to restore their safety. Strategies include soil washing, containment, and bioremediation to reduce environmental damage and enable future land use.

Chapter 6: Spill Control, Confinement, and Decontamination

(Operations Level)

Spill control, confinement, and decontamination are essential practices in managing hazardous material incidents. Effectively addressing these elements is critical to minimizing risks and ensuring safety for first responders, the public, and the environment. When faced with chemical spills or other hazardous situations, understanding how to contain and manage these substances can prevent larger-scale consequences that might arise from uncontrolled releases. The role of those tasked with implementing these strategies involves not only initial awareness and prompt action but also an ongoing commitment to follow established procedures and improve upon them as necessary. This chapter delves into the various techniques used to control spills and confine hazardous materials. It will explore the roles and responsibilities of different personnel involved in these operations, from operations-level responders who initiate containment measures to HazMat technicians with specialized expertise. Readers will gain insight into the methods used to limit exposure and prevent contamination, including the strategic employment of physical barriers and sorbents. In addition to outlining effective response strategies, the chapter examines the importance of coordination and training among responders, ensuring that every individual plays a part in the successful management of hazardous incidents. By understanding these principles, students, firefighters, and first responders will be better prepared to handle potentially dangerous scenarios safely and effectively.

Introduction to Spill Control, Confinement, and Decontamination

Spill control and decontamination are fundamental components in managing hazardous material incidents. They play a critical role in preventing the escalation of these incidents, which can otherwise lead to severe consequences for both people and the environment. Preventing escalation starts with recognizing the inherent dangers of hazardous materials, such as chemical spills or toxic releases that can spread quickly if not contained.

By implementing effective spill control measures, it is possible to limit the spread of hazardous substances. This containment is crucial in protecting both the public and emergency responders from exposure to dangerous chemicals. For instance, prompt response actions, such as establishing evacuation zones and using appropriate protective equipment, significantly reduce health risks. Such measures help ensure that individuals are shielded from direct exposure to harmful substances, thereby maintaining safety and reducing potential injuries or health complications among the affected population.

Furthermore, efficient spill control strategies are vital in minimizing environmental damage. Hazardous substances can contaminate soil and water sources, leading to long-term ecological harm. By containing and mitigating spills promptly, responders can prevent these materials from entering ecosystems and causing widespread environmental degradation. For example, by using diking and damming techniques, the flow of contaminants into water bodies can be halted, thereby protecting aquatic life and preserving water quality.

The success of spill control and decontamination efforts hinges on the roles and responsibilities of different personnel involved in the incident management process. Operations-level responders, HazMat technicians, and Incident Command each have distinct functions that contribute to the overall effectiveness of response strategies.

Operations-level responders often serve on the frontline, executing tasks such as setting up barriers and working to contain spills. Their primary focus is on defensive actions that prevent further contamination or escalation of the incident. As stated in the guidelines, these responders must be well-trained in spill response procedures, equipment use, and protective measures (American Chemical Society, 1995).

HazMat technicians play a more specialized role, possessing the expertise needed to handle and identify various hazardous substances. Their advanced training enables them to assess risks, determine appropriate containment methods, and apply technical solutions for controlling or neutralizing spills. They work closely with operations-level responders to provide guidance and support during the incident. Technicians are also instrumental in evaluating whether the situation requires external assistance from emergency service units or if it can be managed internally (Hazardous Waste Operations and Emergency Response (HAZWOPER) - Standards | Occupational Safety and Health Administration, 2025).

At the top of the chain, Incident Command oversees the entire operation, coordinating efforts and ensuring communication among all parties involved. This role involves making critical decisions on resource allocation, risk assessment, and establishing command post operations. They ensure that all actions align with safety protocols and response plans. The Incident Command function is crucial for maintaining control of the situation and ensuring the safety and efficiency of the response team.

It's essential to recognize the multidimensional nature of spill response and decontamination. Effective management requires careful planning, resource availability, training, and coordination among various levels of personnel. According to the Occupational Safety and Health Administration (OSHA), employers are required to develop comprehensive safety and health programs that incorporate spill response procedures and emergency response protocols. These programs must address the necessary training for responders and ensure medical surveillance to monitor the health of those involved in cleanup operations (Hazardous Waste Operations and Emergency Response (HAZWOPER) - Standards | Occupational Safety and Health Administration, 2025).

Furthermore, it is important for facilities to handle hazardous materials to regularly inspect their storage areas and transport equipment to minimize the risk of spills. This includes maintaining sturdy shelves, implementing secure transport practices, and providing appropriate protective gear for personnel. Regular inspections help in identifying potential hazards before they result in accidental releases, thereby reducing the likelihood of incidents occurring in the first place.

Guidelines for effective spill response include having readily available safety equipment and spill cleanup materials. Facilities must ensure that responders are properly trained in the use of this equipment and that these tools are maintained in good working order. It is equally essential to establish clear procedures for waste disposal following a spill to prevent further contamination. By adhering to these guidelines, responders can take swift and safe action during hazardous events (American Chemical Society, 1995).

Spill Control Techniques

In managing hazardous material incidents, effectively containing and confining spills is crucial to prevent environmental and health hazards. While these terms are often used interchangeably, they represent different aspects of spill management. Containment involves stopping further spread of the hazardous substance, whereas confinement focuses on limiting the area impacted by the spill.

For liquid spills, techniques such as diking, damming, and diverting are essential. Diking refers to creating a barrier around a spill area to prevent the liquid from spreading. This can be achieved using materials like soil, sandbags, or inflatable barriers. Diking is particularly useful for protecting specific areas such as storm drains from contamination (Fletcher, 2023). In contrast, damming involves physical obstructions to control the flow of liquids in waterways. There are three types of dams used for this purpose: static, underflow, and overflow dams. Static dams are ideal for products miscible in water, while underflow and overflow dams are suited for scenarios where either product buoyancy or density relative to water is considered (<i>Document Display (PURL) | NSCEP | US EPA</i>, 2025).

The surface upon which a spill occurs significantly influences how containment and confinement measures are applied. For instance, on paved surfaces, responders may use absorbent materials like pads or booms to quickly contain the spill. These materials are simple to deploy, though their effectiveness can be reduced by wind or the need for proximity to the spill site. On soils, building a dike with available earth materials might be necessary to control the spread of the substance. In waterways, constructing appropriate dams helps manage flow and prevent downstream contamination (Fletcher, 2023).

Guidelines for controlling HazMat releases on various surfaces emphasize that on pavements, quick deployment of absorbents is vital to limit the spread, whereas on soils, creating barriers with natural materials can effectively slow down or redirect the spill. Waterways require strategic placement of dams to maintain control over the substance.

Understanding when to employ defensive versus offensive actions is critical in spill control operations. Defensive actions are preventive, focusing on containing the spill from a safe distance without directly interacting with the hazardous material. Techniques like using sorbents or constructing barriers fall into this category. Offensively, responders might engage more directly, especially in scenarios demanding immediate intervention to neutralize or remove a threat.

The role of adsorption and absorption should also be addressed in the context of spill management. Absorption involves soaking up the spilled material into porous substances, typically resulting in the expansion of the material. Common absorptive materials include clay and vermiculite, which are effective at capturing spills on solid surfaces. Adsorption, on the other hand, involves the adherence of molecules onto the surface of another material. Materials like activated charcoal are particularly useful for adsorbing hazardous vapors present in spill sites (<i>Document Display (PURL) | NSCEP | US EPA</i>, 2025).

In choosing techniques for spill control, it's essential to consider the nature of the surface and the type of hazardous material involved. For liquid chemical spills on pavement, deploying hydrophobic booms may prevent the spread of oil-based contaminants. Conversely, in the presence of corrosive substances, universal or specific hazmat sorbents should be used to ensure compatibility and safety (Fletcher, 2023).

Diversion techniques play a pivotal role in spill management, offering pathways for rerouting hazardous materials away from sensitive areas. This involves creating channels or barriers to guide the flow away from critical infrastructure or ecological zones. By employing charged fire hoses, sandbags, and natural topographical features, first responders can direct the flows toward safer containment areas. Such actions not only protect vulnerable sites but also facilitate subsequent cleanup efforts.

It's important to ensure proper training and preparedness among responders to efficiently implement these strategies. Familiarity with the tools and materials available, as well as understanding each technique's limitations and appropriate applications, enhances response effectiveness. Regular drills and updates on best practices contribute to improved readiness in handling real-world spill scenarios.

Vapor Suppression and Absorption

Managing vapor releases from hazardous materials is a critical component of spill control and decontamination operations. Understanding the risks associated with hazardous vapors, employing effective suppression methods, and selecting appropriate absorbent materials can significantly mitigate potential dangers.

Hazardous vapors pose significant risks to both first responders and the surrounding community. These vapors can be toxic, flammable, or reactive, increasing the likelihood of adverse health effects and secondary incidents (Fletcher, 2023). Influencing factors such as temperature, pressure, and wind speed can affect the dispersion of vapors, complicating response efforts. For instance, warm temperatures and high winds can accelerate the spread of vapors, requiring rapid intervention to prevent exposure.

One essential strategy for managing vapor releases is the use of foam applications and water fogging as suppression methods. Foam blankets are particularly effective in confining hazardous vapors arising from liquid spills. By applying a compatible foam concentrate mixed with water, responders can create a stable foam layer over the spill's surface. This approach reduces vapor emissions by minimizing the exposed surface area of volatile substances (Fletcher, 2023). The application of a foam blanket must be carefully monitored and replenished as needed, as degradation over time can lead to renewed vapor release.

Water fogging, on the other hand, involves the use of a fog nozzle to disperse vapors from an undesired area. Similar to hydraulic ventilation techniques used in firefighting, water fogging acts by moving the vapors away to safer zones until ignition sources can be controlled or removed (Fletcher, 2023). However, it's crucial to understand that some vapors may react with water to form new compounds. For example, anhydrous ammonia gas reacts with water to produce ammonium hydroxide, requiring additional measures to manage any resulting runoff.

Incorporating both absorption and adsorption techniques enhances the effectiveness of managing hazardous vapor releases. Although often confused, these processes are distinct. Absorption involves the uptake of vapors into a medium, much like a sponge soaking up water. Materials such as clay or vermiculite expand as they absorb substances, making them suitable for large volumes of vapor (Fletcher, 2023). Conversely, adsorption captures vapors on the surface of a solid adsorbent like activated carbon. This process occurs at the molecular level, and while expansion is minimal, activated carbon can effectively trap contaminants due to its vast surface area.

Choosing the appropriate material for vapor management is critical. For example, activated carbon is highly effective for adsorption due to its porous nature, offering extensive surface contact for vapor molecules (<i>2.2. Incidents on Land</i>, 2022). In contrast, materials like sand may offer limited absorption capabilities but can provide substantial liquid barrier support in certain scenarios. If rapid deployment is necessary, prefilled sandbags or clay-based products might offer quick solutions in constructing temporary containment structures around spills.

When determining the most suitable technique or material, responders must consider environmental conditions, the chemical properties of the spilled substance, and the availability of resources. Ensuring that responding teams are adequately trained in these strategies will enhance the safety and effectiveness of vapor release management efforts. Training should emphasize understanding material compatibility and proper application methods to avoid exacerbating the situation or causing additional hazards.

Finally, it is crucial for all personnel involved in hazmat response efforts to maintain compliance with established guidelines and protocols. This includes using personal protective equipment (PPE) correctly and remaining aware of their surroundings. Adequate communication among team members and with external partners, such as local authorities and environmental agencies, plays a vital role in coordinating an efficient response.

Emergency Decontamination Procedures

Decontamination is a crucial procedure in the context of hazardous material incidents. Its implementation effectively prevents cross-contamination, significantly reducing the risk of harm to both responders and the general public. Understanding the importance of decontamination underpins our capacity to manage emergencies effectively. Cross-contamination can exacerbate an already hazardous situation, spreading dangerous substances beyond the initial incident site if not properly addressed. Thus, the role of decontamination extends beyond individual safety, encompassing broader community health and environmental protection.

There are two main types of decontamination: emergency and technical approaches. Emergency decontamination involves quick, on-the-spot measures to alleviate immediate threats posed by contaminants, often without full knowledge of the substances involved. It focuses on removing the bulk of contaminants swiftly to protect life and health. On the other hand, technical decontamination is more methodical and detailed, conducted in controlled environments with specialized equipment, usually after identifying the specific hazard present. This approach aims for thorough cleansing, ensuring no harmful residue remains.

The decontamination process for first responders involves clear, systematic steps to ensure efficiency and safety. Initially, responders should remove personal protective equipment (PPE) carefully to prevent self-exposure. PPE serves as the primary barrier between the responder and the hazard; however, its removal is a delicate process that requires adherence to established protocols to avoid transferring contaminants from the gear to the skin or clothing underneath. After PPE removal, responders should proceed through decontamination stations designed to cleanse any potential residual contamination, ensuring their own safety before transitioning back to non-contaminated areas.

Prioritization is key when managing victim and public decontamination during a hazardous materials incident. Victims suffering from life-threatening conditions due to exposure must receive priority treatment to stabilize their condition. Following this, comprehensive decontamination can take place once they are stable. Public decontamination involves communicating clearly and effectively with those involved, providing instructions to minimize panic and confusion. Establishing decontamination corridors helps streamline the process, ensuring everyone exposed receives necessary attention without overwhelming resources.

The sequence typically begins with gross decontamination, essentially washing off as much contaminant as possible quickly. This step often employs water or special solutions to dilute and remove surface contaminants, especially useful in mitigating acute exposure risks. Depending on the nature of the incident, a more rigorous secondary decontamination might follow, incorporating additional cleaning agents suitable for the specific hazards identified, aimed at thorough cleansing.

Effective decontamination also necessitates proper disposal of contaminated materials. Decontaminated articles such as clothing and equipment often still contain residual hazardous substances and must be handled as regulated waste. This process ensures contaminants do not re-enter the environment or pose future risks. All waste generated must comply with local and federal regulations concerning hazardous waste management, involving secure containment and transportation to designated disposal facilities.

To evaluate and enhance the effectiveness of decontamination procedures, regular assessments are imperative. Monitoring involves reviewing each incident's outcomes to identify strengths and areas for improvement in decontamination practices. It ensures continuous adaptation to emerging challenges and advances in decontamination technologies and methodologies.

Summary and Reflections

This chapter explores effective methods and strategies for managing hazardous material incidents, focusing on spill control, confinement, and decontamination. By recognizing the inherent dangers associated with hazardous substances, responders can implement measures to prevent escalation, safeguard public health, and protect the environment. The chapter detailed various techniques such as diking, damming, and diverting, emphasizing their application across different surfaces to contain spills successfully. Additionally, understanding both defensive and offensive actions is crucial, enabling responders to choose appropriate methods based on the circumstances. Furthermore, the integration of absorption and adsorption processes plays a vital role in managing vapor releases effectively, highlighting the importance of selecting suitable materials and techniques.

Responsibility and coordination among personnel at different levels are essential components of effective incident management. Operations-level responders, HazMat technicians, and Incident Command must collaborate to ensure all measures align with safety protocols and resource allocation strategies. Moreover, adherence to established guidelines and protocols fortifies the safety of not only the response teams but also the affected communities and ecosystems. Regular inspections, training, and preparedness initiatives are fundamental to improving the overall efficacy of the response efforts. Through understanding these principles and implementing comprehensive safety programs, facilities handling hazardous materials can significantly reduce risks, ensuring a safe and efficient approach to spill management and decontamination.

20 Questions and Answers Based on Chapter 6: Spill Control, Confinement, and Decontamination (Operations Level)

What is the main goal of spill control and decontamination?

Answer: The main goal is to prevent the escalation of hazardous material incidents, minimize risks to responders, the public, and the environment, and reduce contamination.

What are the roles of operations-level responders in spill control?

Answer: Operations-level responders are responsible for executing containment measures, setting up barriers, and taking defensive actions to prevent the spread of hazardous materials.

How do spill control measures help protect the environment?

Answer: By containing and mitigating spills, responders prevent hazardous materials from contaminating soil, water sources, and ecosystems, thereby reducing long-term environmental damage.

What techniques are used for liquid spill containment?

Answer: Techniques such as diking, damming, and diverting are commonly used to contain and limit the spread of liquid spills.

What is the difference between containment and confinement in spill control?

Answer: Containment stops the spread of the hazardous substance, while confinement limits the affected area of the spill.

What materials are commonly used for diking in liquid spills?

Answer: Materials like soil, sandbags, or inflatable barriers are commonly used to create dikes and contain liquid spills.

What is the role of HazMat technicians in spill control?

Answer: HazMat technicians provide specialized expertise, assess risks, determine containment methods, and apply technical solutions to control or neutralize spills.

How do responders address spills on paved surfaces?

Answer: On paved surfaces, responders often use absorbent materials like pads or booms to quickly contain the spill.

How are hazardous vapors managed during spill control?

Answer: Hazardous vapors are managed through vapor suppression techniques such as foam applications and water fogging to reduce vapor emissions.

What is the difference between absorption and adsorption in vapor management?

Answer: Absorption involves soaking up vapors into a medium (e.g., clay), while adsorption involves the vapor adhering to the surface of an absorbent material (e.g., activated carbon).

How do diversion techniques help in spill control?

Answer: Diversion techniques use channels or barriers to reroute hazardous materials away from sensitive areas, protecting infrastructure and ecosystems.

What is the importance of personal protective equipment (PPE) in spill control?

Answer: PPE is critical in shielding responders from exposure to hazardous materials, preventing contamination during spill response operations.

What are the two main types of decontamination procedures?

Answer: The two main types are emergency decontamination, which is quick and designed to alleviate immediate threats, and technical decontamination, which is thorough and detailed, conducted in controlled environments.

How is gross decontamination performed?

Answer: Gross decontamination involves washing off as much contaminant as possible using water or special solutions, typically to address immediate exposure risks.

Why is proper disposal of contaminated materials important in decontamination?

Answer: Proper disposal ensures that contaminated materials do not re-enter the environment, preventing further contamination and health risks.

What role does Incident Command play during a hazardous materials incident?

Answer: Incident Command oversees the entire operation, coordinates resources, makes critical decisions, and ensures safety protocols are followed.

Why is regular training important for responders in spill control and decontamination?

Answer: Regular training ensures responders are familiar with the latest techniques, tools, and procedures, enhancing their ability to manage hazardous material incidents effectively.

How do dams differ in their use for spill control in waterways?

Answer: There are three types of dams—static, underflow, and overflow—each suited for different spill scenarios based on the density and behavior of the hazardous material relative to water.

How do foam applications help in vapor suppression?

Answer: Foam applications create a stable foam blanket over a spill, reducing the exposed surface area of volatile substances and thereby suppressing hazardous vapors.

How do responders prioritize decontamination efforts during a hazardous material incident?

Answer: Victims with life-threatening conditions receive priority decontamination to stabilize their health, followed by comprehensive decontamination for others based on the level of exposure.

Chapter 7: Advanced Hazardous Materials Response

(Technician Level)

Mastering advanced hazardous materials response at the technician level sharpens the expertise of those tasked with confronting hazardous incidents head-on. These highly skilled responders are equipped to handle dangerous situations where rapid and effective action is imperative. They go beyond basic response, engaging in offensive operations that involve entering contaminated zones and directly controlling hazardous material releases. This specialized knowledge distinguishes them from those operating at awareness or operational levels, forging a path toward superior incident management and safety restoration.

The forthcoming chapter delves into the intricacies associated with being a HazMat technician. Readers will explore the skills and training required for technicians who play an integral role in minimizing environmental impact and public health risks during hazardous events. Discussion spans the critical offensive actions employed by technicians, such as source control techniques and preventative measures against leaks. The importance of using specialized equipment and understanding chemical properties for advanced hazard assessment is highlighted, providing readers valuable insights into the tactical and analytical components of a technician's responsibility. Additionally, proper selection and usage of protective clothing and respiratory systems will be examined to ensure responder safety. The chapter serves as both a comprehensive guide and educational resource for those aspiring to excel in hazardous materials response at a technical level.

Introduction to Advanced HazMat Response

Understanding the complexity and demands of training for a hazardous materials (HazMat) technician level is crucial for those involved in handling dangerous substances. This subpoint explores the distinctions between technician-level responses and awareness or operations levels, emphasizes the role of technicians in offensive strategies for incident mitigation, highlights the necessity of a deep chemical understanding for tactical efficacy, and identifies the primary responsibilities undertaken by technicians.

Technician-level HazMat responders are highly specialized personnel with a skill set that sets them apart from those trained at the awareness or operations levels. Awareness-level responders are often the first on the scene and responsible for initial protective actions such as isolating an area or calling for specialist assistance. Operations-level responders take on a hands-on but defensive role, focusing on minimizing the spread of hazardous material without entering the danger zone (SeventhQueen, 2021). In contrast, technician-level responders engage in offensive actions, directly handling and controlling the sources of hazardous releases, which requires a higher degree of training and expertise (bradskramer, 2020).

Technicians are trained to respond aggressively to known or suspected hazardous material releases. They have the authority to enter the Hot Zone—a heavily contaminated area—where they undertake tasks like shutting down leaks and conducting rescues. This capability is critical in mitigating incidents swiftly and effectively, reducing potential harm to people and the environment. By assuming control in these high-risk situations, technicians play a pivotal role in restoring safety and stability during a HazMat response. The offensive nature of their role requires not only proficiency in practical skills but also the ability to make rapid decisions under pressure (SeventhQueen, 2021).

A profound understanding of chemistry and related scientific disciplines is indispensable for HazMat technicians. While they may not be full-fledged scientists, it is expected that they possess robust knowledge of chemistry, biology, and even nuclear physics to ensure effective decision-making during operations. This chemical comprehension allows technicians to conduct advanced risk assessments, analyze hazards, and choose appropriate methods for containment and neutralization. Such expertise can significantly influence the outcomes of hazmat incidents by ensuring correct procedural adherence and efficient use of resources.

Identifying hazardous substances accurately is among the primary responsibilities of a HazMat technician. This task involves using advanced detection, monitoring, and testing equipment to ascertain the scope of an incident. The selection of proper personal protective equipment (PPE) is guided by the identification of specific chemical threats. Technicians must also determine suitable decontamination procedures to protect both themselves and the environment. Effective substance identification is a cornerstone of their role, enabling tailored interventions that address unique characteristics of various hazardous materials (SeventhQueen, 2021).

Guidelines for identifying hazardous substances begin with comprehensive training in the operation of detection tools such as multi-gas detectors and photoionization detectors (PIDs). These instruments provide real-time data, allowing technicians to gauge concentrations of dangerous substances promptly. Additionally, regular maintenance and calibration of such equipment ensure accuracy and reliability during emergencies. A strong understanding of the limitations and capabilities of each detection method guides technicians in selecting the most appropriate tool for each scenario.

Technician-level training goes beyond technical know-how; it encompasses strategic thinking and adaptability. Training programs are designed to simulate real-life scenarios that expose trainees to a variety of challenges encountered in the field. Integrating simulator detector technology into training enhances the learning experience, offering rigorous, repeatable situations without health and safety risks. Through such immersive training, technicians develop the agility required to navigate complex and dynamic environments effectively.

Offensive Actions in HazMat Response

In the realm of hazardous materials response, technicians are tasked with the crucial responsibility of mitigating potentially dangerous situations through advanced techniques and strategies. Controlling hazardous material releases is a critical component of ensuring safety in such incidents. One of the primary methods involves addressing leaks directly at their source. Identifying the source of a leak quickly can prevent further escalation of the hazard. The initial step often involves assessing the type of material and container involved. Techniques such as remote monitoring and visual inspection are essential for gathering necessary information while maintaining a safe distance.

Effective source control requires a systematic approach. It's vital to stabilize the site before any product removal or containment efforts begin. This includes extinguishing fires, controlling ignition sources, and managing spills and leaks (<i>Hazmat Implementing Response (Product Control Part 1) Flashcards</i>, 2025). Once stabilization is achieved, personnel can focus on direct leak management. This may involve employing physical barriers or temporary sealing solutions to prevent ongoing discharge.

One way to control minor leaks efficiently is through various plugging and patching methods tailored to specific container types. These methods require understanding the nature of the container — be it metal, plastic, or composite — and choosing the appropriate materials for repair. Plugs and patches come in diverse forms, from chemical-resistant putty to inflatable devices that can adapt to different shapes. For instance, in cases where small holes or cracks are present, applying a compatible patch can prevent the leak from worsening.

The careful application of these methods ensures that material remains contained within the compromised vessel. It's crucial for responders to be skilled in selecting the right tools and techniques for the job, as improper patching could lead to further complications. Personnel should undergo training to familiarize themselves with various sealants and mechanical devices designed for specific scenarios.

Overpacking and containment represent another layer of protection when dealing with damaged containers. Overpacking involves placing a compromised container into a larger, more robust one to contain any potential leaks. This method is especially useful when the structural integrity of the original container is severely compromised. Properly executed overpacking not only prevents material release but also facilitates easier transport of hazardous materials to a secure location for final disposal or treatment.

Containment strategies extend beyond overpacking. Techniques like diking, damming, and retaining pools are employed to manage spills, particularly for liquid products. These tactics create physical barriers that confine hazardous substances to a designated area, thereby reducing environmental contamination and exposure risks to personnel and property (<i>Document Display (PURL) | NSCEP | US EPA</i>, 2025).

Mitigating risks from pressurized container leaks presents unique challenges due to the high energy involved. Special valve kits are essential for handling these situations safely. These kits allow technicians to isolate and control the flow of gases or liquids under pressure, effectively preventing further leakage. In some cases, cooling measures might be necessary to reduce the internal temperature of a pressurized tank if heat is contributing to the risk of rupture.

Training and preparedness are integral to effectively utilizing these offensive techniques. Responders must be adept in identifying suitable tactics based on incident characteristics and possess the confidence to implement them swiftly. Frequent drills and simulations enhance the readiness of teams, equipping them with the skills needed to tackle real-world emergencies.

Specialized Tools and Equipment for HazMat Response

In the specialized field of hazardous materials response, HazMat technicians must be equipped with the right tools and equipment to ensure effective and safe operation during incidents. Understanding the function and importance of these tools is vital for technicians who operate in high-stakes environments where precision, safety, and efficacy are paramount.

Chemical detection and monitoring equipment are at the heart of HazMat operations. Multi-gas detectors and Photoionization Detectors (PIDs) play a crucial role in identifying and quantifying hazardous substances in the atmosphere. Multi-gas detectors are designed to detect oxygen levels, combustible gases, and specific toxic hazards such as hydrogen sulfide and carbon monoxide. This ensures that the environment is continuously monitored for any fluctuations or dangerous conditions that could jeopardize safety. PIDs, on the other hand, are used to detect volatile organic compounds (VOCs). This is particularly important in situations involving chemical spills, as PIDs provide real-time data and help technicians assess potential health risks and environmental impacts. The ability to swiftly detect and analyze chemical presence allows technicians to make informed decisions on appropriate containment and mitigation strategies, reinforcing the importance of these instruments in maintaining site safety and compliance with OSHA standards (<i>29 CFR Part 1910 Subpart H—Hazardous Materials</i>, 2015).

Beyond detection, leak control tools form an essential part of the HazMat technician's arsenal. These include non-sparking tools used in environments with flammable gases or vapors. Non-sparking tools, often made from materials such as brass, bronze, or copper-titanium, minimize the risk of sparks that could ignite volatile gases, thus providing an additional layer of safety in explosive environments. Pneumatic and hydraulic devices offer versatile solutions for controlling leaks by applying pressure or creating seals around compromised containers or pipes. Hydraulic tools especially provide the force necessary for clamping, sealing, or cutting through materials without generating heat or static, which could exacerbate the hazard. Guidelines for using overpack drums effectively are critical here, as they allow for the safe containment of leaking barrels or containers. Technicians need to ensure proper procedures are followed to secure the source of the leak and prevent further contamination or exposure.

Specialized containment tools are indispensable when entering confined spaces, where hazardous atmospheres may develop quickly due to limited ventilation. These environments require careful planning and equipment selection to safeguard responders. Devices designed to stabilize atmospheric conditions, such as portable fans and ventilators, are often deployed to reduce concentrations of hazardous gases. Moreover, entry barriers, sealants, and inflatable plugs can help isolate and contain the threat within the confined space. Ensuring all equipment is compatible with the potential chemical hazards is a priority, as incompatibility could lead to adverse reactions.

The use of foam and neutralizing agents serves as a frontline defense in spill mitigation, allowing for immediate control and reduction of chemical hazards. Firefighting foams, such as aqueous film-forming foam (AFFF), are commonly used to create a blanket over flammable liquid spills, suppressing vapor release and preventing ignition. Foam application is particularly effective on hydrocarbon fuel spills, rapidly reducing the fire risk and facilitating cleanup. In contrast, neutralizing agents are employed to counteract corrosive substances, rendering them harmless and preventing damage to infrastructure or harm to personnel. Selecting the appropriate agent involves assessing the chemical properties of the spill and ensuring compatibility with the available resources and environmental considerations.

Chemical Protective Clothing (CPC) and Respiratory Protection

Protective gear selection and usage in hazardous materials (HazMat) operations is critical to ensuring the safety of responders and minimizing risks. Understanding the levels of Chemical Protective Clothing (CPC) and the appropriate use of respiratory protection systems are essential components of this task.

The Environmental Protection Agency (EPA) classifies CPC into four levels—A, B, C, and D—each designed for specific hazard environments and exposure risks. Level A offers the highest level of protection and is used when hazardous substances pose significant respiratory, skin, and eye hazards. This level typically involves a fully encapsulated chemical suit with pressure-demand, full-face Self-Contained Breathing Apparatus (SCBA), inner and outer chemical-resistant gloves, and chemical-resistant boots (https://www.facebook.com/profile.php?id=1091021082, 2019). It is critical in situations where chemicals have high toxicity or carcinogenic properties, such as entering confined spaces with known or potential contaminants.

Level B provides the same respiratory protection as Level A but focuses on less stringent skin protection. It is suitable when liquids are a concern rather than vapors. The equipment includes a liquid splash-protective suit with SCBA. Level B is often used during initial site assessments when understanding the full extent of exposure risks (OSHA Technical Manual (OTM) - Section VIII: Chapter 1 | Occupational Safety and Health Administration, 2025).

Level C protection involves the use of air-purifying respirators (APRs) instead of SCBAs, combined with hooded chemical-resistant clothing that offers similar skin protection as Level B. This level is utilized when contaminants are below OSHA Permissible Exposure Limits (PELs) and can be effectively filtered by the respirator canisters. It is frequently chosen by HazMat workers due to its balance of safety and practicality, particularly when the risk of vapor exposure is low and containment is well-characterized.

Level D provides minimal protection and is primarily for environments where the atmosphere contains no identifiable hazards. It includes basic coveralls, safety boots, and goggles but lacks respiratory protection and is inappropriate for high-risk areas.

In addition to selecting the appropriate CPC level, proper respiratory protection system selection is paramount. SCBAs are vital in environments with unidentified contaminants or oxygen-deficient atmospheres, providing an independent air supply. Conversely, APRs are suited for situations where the air contaminant level is established and filterable, offering more mobility and comfort for the user.

Donning and doffing procedures for CPC and respiratory equipment are integral to maintaining protection integrity. These procedures must be executed meticulously to prevent contamination. Proper donning involves careful inspection of all equipment to ensure functionality, followed by methodical dressing starting with inner garments progressing to outer layers. Effective sealing and fit checks for masks and suits are necessary to prevent exposure. During doffing, special care should be taken to avoid contact with contaminated surfaces of the PPE. A systematic removal process, involving assistance if needed, ensures decontamination.

Decontamination protocols following exposure are crucial to protecting personnel from residual contamination. These protocols entail removing and cleaning contaminated clothing and equipment before leaving the hot zone, thereby reducing exposure risk and preventing cross-contamination. Personnel should follow structured steps, often involving water and detergent washes, followed by specialized neutralizing agents when applicable. Facilities must ensure these protocols are adhered to strictly and consistently implemented.

Final Insights

This chapter has explored the advanced skills necessary for hazardous materials technicians, including offensive strategies essential for efficiently managing dangerous incidents. Technicians distinguish themselves through their capability to engage directly with hazardous releases, requiring a comprehensive understanding of chemistry and tactical procedures. Their training equips them to not only assess and identify chemical threats using sophisticated tools but also to undertake effective containment actions that minimize risks to public safety and the environment.

The practical application of these skills involves hands-on interaction with specialized equipment and execution of strategic leak control and containment methods. By utilizing protective clothing and respiratory gear appropriately, technicians ensure maximum safety during operations. Through rigorous and realistic training scenarios, they develop the resilience and adaptability needed to address complex HazMat challenges swiftly and confidently. This chapter underscores the importance of these competencies in bolstering the readiness and effectiveness of those at the forefront of hazardous material response.

20 Questions and Answers based on Chapter 7: Advanced Hazardous Materials Response (Technician Level)

What distinguishes technician-level HazMat responders from awareness and operations-level responders?

Answer: Technician-level responders engage in offensive actions, entering the Hot Zone to directly control hazardous material releases, while awareness-level responders focus on initial protective actions and operations-level responders handle defensive strategies without entering the danger zone.

What is the primary responsibility of a HazMat technician?

Answer: A HazMat technician's primary responsibility is to directly handle hazardous material releases, such as controlling leaks and mitigating environmental and public health risks.

What are some offensive actions taken by HazMat technicians during an incident?

Answer: Offensive actions include entering the Hot Zone to shut down leaks, conducting rescues, and using techniques such as plugging, patching, overpacking, and containment to control hazardous material releases.

Why is an in-depth understanding of chemistry important for HazMat technicians?

Answer: A strong understanding of chemistry allows technicians to conduct advanced risk assessments, choose appropriate containment strategies, and ensure the safe handling of hazardous materials by understanding their chemical properties.

What is the importance of identifying hazardous substances during a HazMat response?

Answer: Accurate identification is essential for selecting the right personal protective equipment (PPE), determining appropriate containment strategies, and implementing decontamination procedures.

How do HazMat technicians identify hazardous substances at the scene?

Answer: Technicians use advanced detection and monitoring equipment such as multi-gas detectors and Photoionization Detectors (PIDs) to assess hazardous materials and quantify their presence in real-time.

What is source control in HazMat response, and why is it critical?

Answer: Source control involves identifying and addressing the source of a hazardous material release. It is critical for preventing further escalation of the hazard and ensuring effective mitigation of risks.

What is the role of overpacking in hazardous materials containment?

Answer: Overpacking involves placing a compromised container into a larger, more robust container to prevent further leaks and facilitate safer transport for disposal or treatment.

What tools are commonly used by HazMat technicians for leak control?

Answer: HazMat technicians use non-sparking tools, hydraulic and pneumatic devices, and overpack drums to control leaks and contain hazardous materials safely.

Why are non-sparking tools important in HazMat operations?

Answer: Non-sparking tools reduce the risk of igniting flammable gases or vapors, providing additional safety when working in explosive environments.

What is the purpose of using foam and neutralizing agents in HazMat response?

Answer: Foam is used to suppress vapor release and prevent ignition in flammable liquid spills, while neutralizing agents counteract corrosive substances, rendering them harmless to responders and the environment.

What are the different levels of Chemical Protective Clothing (CPC)?

Answer: The four levels of CPC are A (highest protection), B (similar to A but with less skin protection), C (air-purifying respirators and hooded chemical-resistant clothing), and D (minimal protection for non-hazardous environments).

When should a HazMat technician use Level A CPC?

Answer: Level A is used in situations with significant respiratory, skin, and eye hazards, such as when handling highly toxic or carcinogenic materials, requiring full encapsulation with a Self-Contained Breathing Apparatus (SCBA).

What is the function of a Self-Contained Breathing Apparatus (SCBA) in HazMat response?

Answer: SCBAs provide an independent air supply to HazMat responders, especially in environments with unidentified contaminants or oxygen-deficient atmospheres.

What does decontamination of HazMat PPE involve?

Answer: Decontamination involves removing and cleaning contaminated clothing and equipment, often using water, detergent, and specialized neutralizing agents to prevent exposure and cross-contamination.

What is the significance of using a Photoionization Detector (PID) during a HazMat response?

Answer: A PID is used to detect volatile organic compounds (VOCs) in the atmosphere, providing real-time data to assess health risks and environmental impacts during chemical spills or hazardous incidents.

How does a HazMat technician control a pressurized container leak?

Answer: Technicians use special valve kits to isolate and control the flow of gases or liquids under pressure, potentially employing cooling measures to reduce the risk of rupture.

What is the role of training and simulation in HazMat technician preparedness?

Answer: Training and simulation expose technicians to real-life challenges, helping them develop the skills and adaptability needed to handle dynamic HazMat incidents effectively.

What are the dangers of confined spaces during HazMat incidents, and how can technicians mitigate these risks?

Answer: Confined spaces can lead to hazardous atmospheres quickly developing. HazMat technicians mitigate risks by using equipment like portable fans, ventilators, and sealants to stabilize atmospheric conditions.

Why is proper donning and doffing of CPC and respiratory protection crucial?

Answer: Proper donning and doffing ensure that responders remain protected from contaminants, prevent exposure during the removal of gear, and maintain the integrity of PPE throughout the operation.

Chapter 8: Hazardous Materials Incident Command and Risk Assessment

(Technician Level)

Managing hazardous materials incidents requires a structured approach that integrates incident command and risk assessment at the technician level. Coordinating such responses involves defining clear roles and responsibilities within an established command structure, which ensures efficiency and improves safety for everyone involved. The presence of qualified technicians is crucial in this setting, as their expertise in assessing environmental conditions and potential threats serves as the backbone for informed decision-making. Their role extends beyond mere data collection; they actively contribute to the formulation and execution of strategic response plans known as Incident Action Plans (IAPs). These technicians provide the necessary information and analysis that allow Incident Commanders to make tactical decisions aimed at minimizing risks during hazardous materials incidents.

This chapter delves into various aspects central to effective response management in hazardous materials incidents, focusing on structured leadership and continuous risk analysis. It starts by exploring how structured command systems and adherence to regulatory mandates from organizations like OSHA, NFPA, and FEMA contribute to public safety and emergency preparedness. The discussion moves into risk assessment techniques, emphasizing the importance of ongoing situational analyses and hazard evaluations conducted by HazMat technicians. This continuous risk analysis enables dynamic adaptations to evolving circumstances, ensuring that strategies remain aligned with real-world demands. Furthermore, the chapter outlines the integral role of communication in maintaining a seamless flow of information, which is vital for implementing effective IAPs. By covering these critical elements, the chapter provides comprehensive insights into how a combination of structured command and proactive risk management enhances the overall effectiveness of managing hazardous materials incidents.

Introduction to Incident Command and Risk Assessment in HazMat Incidents

In the realm of hazardous materials (HazMat) incidents, structured leadership shapes the effectiveness of emergency response. At the core of managing these emergencies, particularly for HazMat Technicians and first responders, lies the importance of a well-established command structure. Structured leadership is not just about giving orders; it is central to ensuring everyone involved understands their roles and responsibilities, translating into a more efficient and safer operation.

Structured leadership is crucial when coordinating complex HazMat emergencies as it provides clarity in chaotic situations. In environments where seconds count and precision is vital, having a definitive command structure allows for streamlined communication and decision-making. This ensures that all actions are coordinated effectively, providing safety for both responders and the public. The clearer the structure, the smoother the operation, leading to reduced risks and heightened safety measures.

A HazMat Technician's role is integral in this context, especially in gathering and analyzing critical information necessary for sustained decision-making during an incident. These technicians offer the backbone of data collection, verifying the nature and scope of the threat through environmental assessments and monitoring potential hazards. They bear the responsibility of informing the Incident Commander (IC), which allows for informed tactical decisions. Furthermore, their involvement in developing and executing the

Incident Action Plan (IAP) is pivotal. The IAP serves as the strategic blueprint for the incident response, outlining objectives, resources, and recovery efforts. By contributing accurate field data and hazard analyses, technicians ensure the IAP is both relevant and adaptable, catering to evolving incident dynamics.

The role of structured command transcends individual organizations, aligning with mandates from key regulatory bodies like the Occupational Safety and Health Administration (OSHA), National Fire Protection Association (NFPA), and Federal Emergency Management Agency (FEMA). These organizations emphasize the necessity of a structured command system through legal requirements and guidelines, illustrating its significance in public safety. OSHA's standards, such as 29 CFR 1910.38 and 1910.120, lay out clear requirements for emergency planning and response, emphasizing the need for a unified command structure (<i>Emergency Preparedness and Response: Getting Started | Occupational Safety and Health Administration</i>, 2015).

Continuous risk analysis further complements structured leadership by addressing the dynamic nature of hazardous material incidents. Monitoring changing conditions allows responders to adapt their strategies and remain in control over the incident. Flexible decision-making hinges on a continuous flow of situational assessments, empowering teams to modify tactics in real-time. Risk analysis is ongoing, ensuring that each phase of the response is adjusted as new information emerges or as conditions change. This proactive approach minimizes unexpected challenges and keeps the overarching strategy aligned with real-world demands.

Gathering critical incident information holds significant weight in shaping effective responses. For HazMat technicians, conducting thorough site assessments and hazard evaluations is not just part of the process but is foundational to strategic response. Utilizing advanced tools and technologies, technicians collect data on chemical properties, potential exposure risks, and environmental factors. This information informs the IAP, offering a comprehensive picture of the scene, thereby aiding in the crafting of precise response strategies tailored to the unique challenges of each incident. Through detailed observation and documentation, technicians can identify shifts in threat levels or emerging hazards, leading to timely updates to the incident plan.

Structured leadership also requires adherence to guidelines provided by prominent organizations concerned with public and occupational safety. The NFPA's codes, for example, offer frameworks intended to minimize risks related to fire and other hazards, highlighting the importance of competent command systems in volatile situations. Similarly, compliance with FEMA's directives ensures uniformity and efficiency across multiple agencies during large-scale incidents, promoting effective interagency collaboration. Such collaborations leverage collective resources and expertise, optimizing response efforts while upholding regulatory expectations. Moreover, these guidelines serve as benchmarks, fostering consistency and reliability in emergency management.

In maintaining effective control over incidents, continuous risk assessment maps onto strategic and leadership roles. Being able to anticipate potential repercussions or shifts in the situation enhances incident management capabilities. Regular training on emergent threats and hands-on simulations can further hone responders' ability to apply theoretical knowledge to practical scenarios, thus reinforcing the concept of adaptive management.

Establishing an Incident Action Plan (IAP)

Creating and executing an effective Incident Action Plan (IAP) for hazardous materials (HazMat) incidents is crucial in safeguarding the lives of both responders and the public while minimizing environmental impact. The IAP serves as a roadmap, outlining strategies and objectives to manage and control incidents safely and efficiently.

Defining incident objectives and operational priorities is paramount in developing a cohesive response strategy. An effective IAP begins with clear, measurable goals that align with the overall mission of the incident response. These objectives provide a sense of direction and help prioritize actions during the often-chaotic environment of HazMat incidents. For instance, ensuring the safety of all personnel is always a top priority, followed by stabilizing the incident to prevent further harm. Operational priorities might include the protection of critical infrastructure, containment of hazardous substances, and mitigation of environmental damage. Defining these elements clearly enables responders to focus their efforts on actions that directly support the primary aims of the operation (2024).

An essential aspect of creating a robust IAP is gathering and analyzing incident information. This involves assessing environmental conditions such as weather patterns and topography, which can significantly influence the spread of hazardous materials. For example, wind direction and speed could affect the dissemination of airborne toxins, necessitating the establishment of safe zones upwind of the incident site. Additionally, understanding the available resources, including personnel, equipment, and the capacities of nearby facilities, informs decision-making and resource allocation. This comprehensive analysis aids in predicting potential incident developments and crafting contingency plans to address unforeseen challenges (2024).

Once incident objectives are defined and relevant data collected, developing tactical response strategies becomes the next critical step. Setting containment and mitigation objectives is essential to limit the spread of hazardous substances and reduce their impact. This may involve deploying barriers or using absorbent materials to control spills or leaks. Establishing safety zones ensures that responders work within areas deemed safe from exposure, minimizing the risk of injury or contamination. Coordinating decontamination procedures is another vital component, ensuring that all personnel and equipment are thoroughly cleaned before leaving the incident site, thus preventing the spread of contaminants beyond the affected area.

Evacuation procedures must also be carefully planned and communicated. This includes determining who needs to evacuate, how they will be evacuated, and where they should go. Clear instructions and designated evacuation routes are essential to avoid confusion during high-pressure situations. Effective coordination with local agencies, such as law enforcement and emergency medical services, enhances the efficiency of these operations, helping to safeguard public health and safety (<i>5 Keys to a Great Fireground Incident Action Plan</i>, 2017).

Communication plays a pivotal role in the successful execution of an IAP. Regular briefings ensure that all responders are fully informed of the plan's details and any updates as the situation evolves. These briefings provide an opportunity to clarify roles and responsibilities, address concerns, and reinforce safety protocols. Utilizing standardized communication protocols, including radios and predefined channels, helps maintain clear and concise exchanges between individuals and teams on the ground. This reduces the likelihood of misunderstandings that could lead to errors in executing the plan.

Moreover, flexibility is key in managing HazMat incidents effectively. The dynamic nature of these situations means that initial plans may need adjustments as new information emerges or conditions change. Having a structured framework in place allows for quick adaptations to tactics without losing sight of the overarching objectives. For instance, if shifting weather conditions threaten to expand the impacted area, adjusting boundaries for safety zones or modifying evacuation plans might be necessary. Maintaining open lines of communication ensures that all team members stay informed about such changes, enabling them to act swiftly and efficiently (<i>5 Keys to a Great Fireground Incident Action Plan</i>, 2017).

Risk Assessment and Hazard Analysis

To effectively manage hazardous materials (HazMat) incidents, a systematic approach to identifying, evaluating, and managing risks is crucial. The first step in this process involves identifying physical, health, and environmental hazards. Understanding these hazards is foundational for conducting a comprehensive hazard analysis, which lays the groundwork for subsequent risk management efforts.

Physical hazards often include fire, explosion, and reactivity risks. Health hazards might involve exposure to toxic substances, corrosive materials, or radiation. Environmental hazards could encompass water contamination from chemical spills or air pollution due to noxious fumes. Each of these hazard types must be thoroughly assessed to understand their potential impact on incident responders, the public, and the environment. Identifying these hazards requires gathering data from various sources such as Safety Data Sheets (SDS), past incident reports, and real-time observations at the incident site.

Once hazards are identified, the focus shifts to risk-based decision-making, which relies heavily on using hazard assessment tools. These tools help establish exposure limits and determine appropriate response strategies. The National Institute for Occupational Safety and Health (NIOSH) and the Environmental Protection Agency (EPA) provide guidelines and tools that can be invaluable in assessing risk levels and developing strategies to mitigate those risks. For instance, air quality monitors and chemical detection equipment can give immediate data regarding the presence and concentration of hazardous substances, informing whether evacuation or containment measures are necessary.

Conducting site characterization and safety evaluations is another critical component of risk assessment. This involves a detailed examination of the incident locale to ascertain immediate and long-term risks to responders and the public. Site characterization may include mapping out hazard zones, identifying the location and type of HazMat involved, and understanding the geographic and meteorological conditions that could influence the spread of contaminants. For example, in an industrial setting, responders would need to evaluate the layout of the plant, access routes, and any nearby residential areas that might be affected by an incident.

Safety evaluations go hand-in-hand with site characterizations, focusing on the practical aspects of ensuring responder safety, such as the availability and condition of personal protective equipment (PPE), communication systems, and emergency procedures. Regular drills and training sessions should be conducted to ensure that all personnel are familiar with emergency protocols and can respond swiftly and safely when a HazMat incident occurs.

Implementing a continuous risk assessment process is essential for monitoring changes in the incident environment and adapting response strategies accordingly. This dynamic approach enhances overall incident management by allowing responders to react proactively to new information or evolving circumstances. Continuous assessment might involve regular updates on weather conditions, shifts in wind patterns that could affect airborne toxins, or changes in the structural integrity of containers housing hazardous materials.

Additionally, documenting all findings and actions taken throughout the risk assessment process is vital. Detailed records not only ensure compliance with regulatory requirements but also serve as valuable resources for post-incident analysis and future training programs. Effective documentation can highlight lessons learned, identify areas for improvement, and contribute to the development of best practices in HazMat incident response.

Engagement with stakeholders, including local authorities, healthcare facilities, and community organizations, is also crucial during the risk assessment process. By involving these groups, incident commanders can gain a more comprehensive view of the potential impacts of a HazMat incident and develop more effective response plans.

Stakeholder collaboration fosters a culture of safety, ensuring everyone is aware of their roles and responsibilities during an emergency.

Incorporating the risk-based approach (RBA) into every phase of HazMat incident management is a strategic way to organize and streamline responses. The RBA integrates the NFPA 470 APIE (analyze, plan, implement, and evaluate) process, providing a structured framework adaptable to various scenarios. This model assists in breaking down complex situations into manageable components, promoting informed decision-making and efficient resource allocation.

The analyze phase begins upon receiving an alarm and continues through the incident's duration. Initial assessments guide the establishment of the Incident Command System, where preliminary scene reports inform subsequent action planning. Upon arrival, responders gather additional information to refine the scope and nature of the hazard. During this stage, leveraging expert judgment and historical data can aid in establishing a baseline understanding of the situation.

Subsequent phases of the APIE model enable a logical progression from analysis to planning, wherein specific objectives are defined, and tactical decisions are made. This sequence ensures that every aspect of the incident is addressed methodically, minimizing risks to responders and the broader community. Plans remain flexible, allowing for adjustments based on real-time developments or emerging threats.

Coordination with Local, State, and Federal Agencies

In the complex landscape of hazardous materials (HazMat) incidents, effective management and resolution hinge upon robust interagency collaboration. This cooperation not only ensures swift action but also guarantees a well-rounded response due to the diverse expertise brought together from various agencies.

Central to this collaborative effort is a thorough understanding of the Incident Command System (ICS). The ICS provides a standardized approach to command, control, and coordination during large-scale emergencies, including HazMat situations. By integrating facilities, equipment, personnel, procedures, and communication within a unified organizational structure, ICS enables incident managers to address key concerns efficiently without overlooking critical aspects of the operation (<i>NIMS Components - Guidance and Tools | FEMA.gov</i>, n.d.). This system's flexibility allows it to be adapted to incidents of any magnitude while ensuring that all responders are on the same page, thus avoiding confusion and duplication of efforts.

Interagency collaboration plays a pivotal role in resource sharing, drawing on local, state, and federal agencies' assets and expertise. This network enhances the overall capability to manage HazMat incidents through shared resources and knowledge. For instance, agencies can pool specialized equipment or tap into expert teams that may not be available within a single entity. Such resource-sharing expedites response times and fortifies the overall incident management strategy (<i>Role of Transportation Management Centers in Emergency Operations Guidebook - Section 3 Pre-Event Activities</i>, n.d.).

Legal and regulatory compliance forms the backbone of interagency responses. Agencies must adhere to standards set forth by bodies such as OSHA, NFPA, EPA, and DOT to ensure safety and legality in operations. Compliance with these regulations not only safeguards responders and the public but also facilitates smoother interactions across agency lines. When all parties adhere to common standards, the groundwork for effective collaboration is laid, reducing misunderstandings and legal liabilities.

A key component of enhancing interagency collaboration is establishing mutual aid agreements and forming regional response teams. Mutual aid agreements provide a formal framework for assistance between communities and states, granting access to additional resources when needed. These agreements, which can also involve private sector entities and NGOs, are crucial in extending the capacity to handle incidents beyond the capabilities of a single jurisdiction (<i>NIMS Components - Guidance and Tools | FEMA.gov</i>, n.d.). Regional response teams further bolster this approach by promoting joint training exercises and pre-incident planning, fostering a spirit of readiness and cohesion among participating units.

These preparations ensure that when an incident occurs, agencies can operate seamlessly, leveraging their collective strengths. Moreover, conducting regular joint exercises helps to identify potential gaps and streamline strategies, improving overall response efficiency.

Real-world examples underscore the importance of interagency collaboration in managing HazMat incidents. During Hurricane Katrina, for example, the lack of coordinated response highlighted the need for a more integrated approach. In contrast, the seamless cooperation witnessed during the Deepwater Horizon oil spill cleanup demonstrated the effectiveness of a well-coordinated interagency operation. Multiple agencies came together, each contributing its expertise, whether in environmental protection, maritime operations, or health assessments, showcasing the power of collaboration in overcoming complex challenges.

Furthermore, incorporating guidelines for detecting and quantifying hazardous atmospheres, as well as identifying oxygen-deficient, flammable, or toxic environments, becomes crucial. These guidelines allow responders to assess risks accurately and take appropriate measures, protecting both themselves and the public. Utilizing real-time data to adjust response tactics can enhance decision-making, ensuring that strategies remain relevant as conditions evolve. Additionally, conducting briefings and maintaining situational awareness promote transparency and informed decision-making among all responding units.

Concluding Thoughts

In this chapter, we explored the critical role of incident command and risk assessment in managing hazardous materials incidents effectively. The discussion highlighted how structured leadership is central to coordinating complex emergencies, providing clarity and streamlined communication necessary for swift decision-making. We delved into the responsibilities of HazMat Technicians, who gather and analyze vital data crucial for informed tactical decisions, thus ensuring a relevant and adaptable Incident Action Plan (IAP). Additionally, we examined the importance of compliance with guidelines from key regulatory bodies such as OSHA, NFPA, and FEMA, which underscore the need for a unified command system. Through ongoing risk analysis, responders are equipped to adapt strategies in real-time, minimizing unexpected challenges and aligning with evolving conditions.

The chapter also underscored the necessity of developing a comprehensive IAP, focusing on clear objectives and operational priorities to guide response efforts. By assessing environmental factors and available resources, responders can develop effective containment and mitigation strategies to safeguard the public and environment. Emphasizing interagency collaboration, the chapter discussed how shared expertise and resources enhance incident management efficiency. Through examples like the Deepwater Horizon oil spill cleanup, the text illustrated the power of coordinated responses in overcoming complex challenges. This chapter aimed to equip students and first responders with an understanding of the pivotal components of emergency management, emphasizing that informed leadership and strategic planning are paramount in effectively responding to HazMat incidents.

20 questions and answers based on the content from Chapter 8: Hazardous Materials Incident Command and Risk Assessment (Technician Level):

Why is structured leadership crucial in managing hazardous materials (HazMat) incidents?

Answer: Structured leadership is crucial because it ensures clarity of roles and responsibilities, facilitates effective communication, and streamlines decision-making, which reduces risks and improves safety for both responders and the public.

What is the role of a HazMat Technician during an incident?

Answer: A HazMat Technician is responsible for gathering and analyzing critical data about environmental conditions and potential hazards, which informs the Incident Commander's tactical decisions and contributes to developing the Incident Action Plan (IAP).

What is an Incident Action Plan (IAP)?

Answer: The IAP is a strategic document that outlines the objectives, resources, and response strategies to manage a HazMat incident safely and effectively, guiding decision-making throughout the incident.

How do regulatory bodies like OSHA, NFPA, and FEMA influence HazMat incident management?

Answer: These regulatory bodies set standards that require a unified command system, providing guidelines for effective incident management to ensure responder safety and public protection during HazMat incidents.

Why is continuous risk analysis important during HazMat incidents?

Answer: Continuous risk analysis allows responders to monitor changing conditions, adapt strategies, and make real-time adjustments, ensuring effective response and minimizing unforeseen challenges.

What type of data do HazMat Technicians gather during an incident?

Answer: HazMat Technicians collect data on chemical properties, potential exposure risks, environmental factors, and other relevant information that informs the Incident Action Plan.

What are some examples of physical hazards in a HazMat incident?

Answer: Physical hazards in a HazMat incident can include fire, explosion, and reactivity risks associated with hazardous materials.

How does wind direction influence HazMat incident response?

Answer: Wind direction can affect the spread of airborne toxins, which is why safe zones must be established upwind to protect responders and the public from exposure.

What are some key components of an effective IAP?

Answer: Key components include defining incident objectives, conducting environmental assessments, developing tactical response strategies, and planning evacuation procedures.

What role does communication play in the execution of an IAP?

Answer: Communication ensures that all responders are informed of the plan, updated as the situation evolves, and able to clarify roles and responsibilities to avoid errors and confusion.

How do HazMat Technicians contribute to formulating the IAP?

Answer: They provide critical information and analyses from the field, allowing Incident Commanders to make informed decisions and develop response strategies tailored to the unique challenges of the incident.

What tools do HazMat Technicians use to assess risk during an incident?

Answer: HazMat Technicians use tools like air quality monitors, chemical detection equipment, and Safety Data Sheets (SDS) to evaluate risks and guide response decisions.

What is the purpose of site characterization in risk assessment?

Answer: Site characterization involves assessing the incident location to identify hazard zones, potential exposure risks, and other environmental factors that influence response strategies.

What is a risk-based approach (RBA) in HazMat incident management?

Answer: The RBA involves analyzing hazards, planning response strategies, implementing them, and evaluating their effectiveness throughout the incident, ensuring that resources are allocated efficiently, and risks are minimized.

How does interagency collaboration contribute to effective HazMat incident response?

Answer: Interagency collaboration enables resource sharing, knowledge exchange, and coordinated efforts, improving overall incident management and response time.

What is the role of mutual aid agreements during HazMat incidents?

Answer: Mutual aid agreements formalize assistance between communities or agencies, providing additional resources and support when needed during a large-scale HazMat incident.

Why is adherence to legal and regulatory compliance important in HazMat responses?

Answer: Compliance ensures that all responding agencies operate safely and within the law, fostering smoother coordination between agencies and minimizing legal risks.

What kind of hazards should be identified in the initial stages of a HazMat incident?

Answer: Initial assessments should identify physical hazards (fire, explosion), health hazards (toxicity, radiation), and environmental hazards (contamination of air or water).

How does the National Institute for Occupational Safety and Health (NIOSH) contribute to risk assessment in HazMat incidents?

Answer: NIOSH provides guidelines and tools for assessing exposure limits, ensuring that response strategies are based on accurate and up-to-date safety standards.

What is the role of stakeholder engagement during a HazMat incident?

Answer: Stakeholder engagement, including coordination with local authorities, healthcare facilities, and community organizations, helps broaden the understanding of the incident's impact, ensuring more effective response planning and decision-making.

Chapter 9: Special Hazards and Terrorism Response

(Technician Level)

Responding to special hazards and terrorism at the technician level demands a sophisticated understanding of hazardous materials incidents. These situations are complex and dynamic, requiring responders to have specialized skills for identifying and managing diverse threats. Special hazards encompass a wide range of dangerous substances, while terrorism adds an unpredictable layer of risk. The intersection of these elements introduces challenges that necessitate advanced training and acute situational awareness. In a world where the nature of threats is continually evolving, it becomes imperative for responders to be proficient in using specialized equipment and applying strategic response protocols effectively.

This chapter delves into the critical aspects of technician-level response to special hazards and terrorism within hazardous materials incidents. It systematically explores various threat categories, including biological, radiological, and chemical hazards, providing insights into the necessary skills and knowledge required for effective management. Readers will gain an understanding of the specific training needed for dealing with these challenges, focusing on detection devices and personal protective equipment tailored to each hazard type. Additionally, the chapter emphasizes the importance of interagency coordination, highlighting how collaboration enhances emergency management efforts. By examining these key components, the chapter equips readers with the essential tools and strategies for navigating the complexities of hazardous material responses and safeguarding public safety.

Introduction to Unique Challenges in HazMat Incidents

In the ever-evolving landscape of hazardous materials (HazMat) incidents, the unique challenges faced by technicians require a deep understanding of various threat categories, including those related to terrorism. Responding to terrorist-related threats is not only about containing and managing immediate dangers but also requires extensive knowledge about biological, radiological, and chemical hazards.

One of the key elements involved in dealing with these threats is understanding the nature and behavior of different agents. Biological threats, such as those posed by anthrax or smallpox, demand a comprehensive grasp of pathogen transmission methods and containment strategies to minimize exposure and spread. Radiological hazards, on the other hand, involve risks from nuclear materials that necessitate familiarity with radiation types—alpha, beta, and gamma rays—and their respective protective measures. Chemical threats are equally diverse, ranging from toxic industrial chemicals to nerve agents, each requiring specific safety protocols and equipment for safe management. The knowledge of these threats underscores why specialized training is paramount for HazMat responders. According to Bland (2013), the complexities of Chemical, Biological, Radiological, and Nuclear (CBRN) agents require responders to be proficient in identification, decontamination, isolation, and hazard management.

The necessity for specialized training cannot be overstated. Emergency responders must undergo rigorous education to effectively identify and neutralize various threats. This training includes the use of detection devices and personal protective equipment (PPE) tailored to specific hazards. For instance, handling chemical agents often involves wearing encapsulated suits that prevent any skin contact. Similarly, responding to radiological threats may require dosimeters to monitor radiation exposure levels, ensuring that personnel do not exceed safe limits while operating in contaminated environments. Moreover, continuous professional development through drills and scenario-based training is essential to keep responder's adept at utilizing advanced technologies and methodologies relevant to current and emerging threats.

Equally important is the deployment of specialized equipment designed to manage complex emergencies. Detection and analysis tools are critical in identifying the presence of hazardous substances, guiding appropriate response actions. Sophisticated sensors can detect minute traces of harmful agents in the environment, allowing responders to implement targeted interventions swiftly. Advanced communication systems also play a vital role in maintaining a seamless flow of information among all units involved in an incident, ensuring that operational strategies are well-coordinated and effective.

The evolving threat landscape further complicates the dynamics of HazMat responses. Global terrorism and environmental changes introduce new challenges that require proactive adaptability from response teams. As terrorists continuously explore novel means to disrupt societies, the threat spectrum expands, with potential scenarios involving unconventional weapons or improvised explosive devices. Moreover, climate change contributes to increased frequencies of natural disasters, compounding the risk factors in hazardous material incidents.

To navigate this shifting terrain, HazMat technicians are indispensable in implementing safety protocols and coordinating efforts among multiple agencies. These professionals serve as the backbone of any response operation, drawing on their expertise to develop and enact strategic plans that prioritize both public and responder safety. Their role encompasses everything from initial threat assessment and zone establishment to casualty management and recovery operations.

Coordination with various agencies, including law enforcement, medical services, and environmental protection agencies, is crucial to ensure a comprehensive response framework. The ability of HazMat teams to work synergistically with other organizations enhances the overall efficacy of emergency management efforts. Interagency collaboration facilitates resource sharing and joint decision-making, streamlining processes and optimizing outcomes.

Moreover, the increasing reliance on technology within the field necessitates a proficient workforce capable of harnessing digital tools to improve response capabilities. Data analytics, for instance, can predict potential threat scenarios based on historical patterns and real-time intelligence, enabling preemptive measures that mitigate risks before they manifest. Remote monitoring systems offer real-time surveillance of sites, providing critical insights that guide tactical operations.

Managing Biological, Radiological, and Chemical Threats

Navigating biological, radiological, and chemical threats at a technician level is essential for handling hazardous materials incidents effectively. Biological agents such as anthrax represent a significant risk due to their ability to spread through multiple exposure routes, including inhalation, ingestion, or skin contact. Understanding these pathways is crucial for responders since it allows them to evaluate the extent of contamination and apply appropriate safety measures (Bland, 2013). For instance, inhalation of anthrax spores can lead to severe respiratory issues, highlighting the importance of protective respiratory equipment.

Radiological hazards come from various nuclear materials that emit radiation, which can be harmful or lethal in high doses. The understanding of different radiation types—alpha, beta, gamma, and neutron—is vital for developing appropriate protective strategies. Each type of radiation requires specific measures; for example, alpha particles can be blocked by clothing or skin, whereas gamma rays require denser materials like lead for protection. Responders need to use radiation detectors to assess environmental risks and decide on personal protective equipment (PPE) levels required to prevent exposure (Kumar et al., 2010). Implementing safety protocols for highly hazardous materials involves proper training to ensure that technicians are proficient in identifying and mitigating radiological threats.

Chemical threats pose unique challenges, particularly with toxic industrial chemicals (TICs) that may be encountered during responses. TICs can include substances like chlorine, ammonia, or hydrogen cyanide, each requiring specialized handling and mitigation techniques. Technicians must be knowledgeable about chemical properties and potential reactions to other substances. Personal protective equipment plays a crucial role in ensuring responder safety while dealing with these chemicals. Guidelines must be adhered to, including maintaining upwind positions and using appropriate decontamination procedures to minimize secondary contamination (Kumar et al., 2010).

Collaboration with Chemical, Biological, Radiological, and Nuclear (CBRN) units is imperative for advanced decontamination and neutralization processes. CBRN units have specialized equipment and expertise that enhances the response capabilities during hazardous material incidents. They provide critical support in identifying unknown agents, conducting thorough decontamination, and ensuring that areas are safe post-incident. Interoperability between local HazMat teams and CBRN units is vital; clear communication and coordinated efforts enable effective management of complex scenarios.

In the presence of a biological threat like anthrax, it is crucial to assess the risk of terrorism-related incidents and specialized threats comprehensively. This includes recognizing unusual patterns of illness or animal death, which could signal a biological attack. Rapid identification and containment can limit the spread and impact of such threats. Effective risk assessment involves not only understanding the nature of the agent but also considering environmental factors and potential targets, enabling responders to make informed decisions on how to proceed (Bland, 2013).

Radiological incident responses require meticulous planning and execution of safety protocols. Establishing exclusion zones and controlling access helps in minimizing exposure to radiation. Educating technicians on dose limits and cumulative exposure can further safeguard their health. In situations involving potential radiological dispersal devices, timely evacuation or shelter-in-place decisions must be made based on comprehensive situational awareness (Kumar et al., 2010). Implementing robust safety protocols emphasizes the importance of preparation and training in scenarios involving dangerous materials.

Coordinating with law enforcement, federal agencies, and specialized HazMat teams ensures a well-rounded approach to managing chemical threats. Given the potential for mass casualties and infrastructure damage, collaboration with these entities provides additional resources, intelligence, and logistic support necessary for an effective response. Federal agencies often bring in advanced analytical capabilities and large-scale decontamination resources that bolster local efforts. Seamless communication channels among these groups enhance their capability to react quickly and efficiently, mitigating the overall impact of the incident.

Illicit Drug Labs and the Opioid Crisis

In the realm of hazardous materials incidents, illicit drug labs pose a unique challenge for responders, demanding a high level of caution and expertise. These clandestine laboratories are often associated with the production of methamphetamines and opioids, such as fentanyl, which present significant fire, explosion, and chemical exposure hazards.

Understanding these dangers is essential for ensuring the safety of both first responders and the communities they serve.

Illicit drug labs frequently house volatile chemicals that can be ignited easily, leading to fires or explosions. Methamphetamine production, for example, involves substances like anhydrous ammonia, hydrogen chloride, and lithium, which are not only highly flammable but also toxic. Responders must be adept at identifying these chemicals to mitigate the risks effectively. This necessitates specialized training in chemical recognition and an understanding of how different substances interact. Equipment used in these labs, such as makeshift distillation apparatuses, increases the risk of accidents due to their often rudimentary and unstable nature.

The opioid crisis, notably involving fentanyl, adds another layer of complexity. Fentanyl is about 50 times more potent than heroin and poses severe health risks even at minimal exposure levels, making it imperative for responders to employ strict protective measures. As outlined by the "FENTANYL: THE next WAVE of the OPIOID CRISIS," just two milligrams of fentanyl can be lethal when inhaled, swallowed, or absorbed through the skin, underscoring the need for personal protective equipment (PPE) to prevent accidental exposure (-FENTANYL: THE next WAVE of the OPIOID CRISIS, n.d.).

Coordination with law enforcement is crucial at these sites, not only to secure the scene but also to ensure compliance with legal protocols. The presence of illegal operations means that securing evidence without compromising the investigation is vital. Law enforcement agencies often have a better grasp on handling criminal elements and can provide the necessary backup to maintain safety and order. Moreover, integrated efforts between various agencies enable rapid decision-making and effective allocation of resources, further enhancing the response team's efficiency.

Addressing fentanyl-related incidents requires awareness of its highly potent nature and the frequent need for higher doses of naloxone to counteract overdoses. A multi-agency approach, as demonstrated in New Haven, involves public health authorities, emergency services, and law enforcement working in unison to identify sources, disseminate public alerts, and provide access to life-saving interventions like naloxone (Tomassoni et al., 2017). Such collaboration ensures a comprehensive strategy, limiting the impact of outbreaks by facilitating rapid response and recovery.

To handle these complex environments safely, guidelines are essential for protecting personnel and managing the varying threats encountered. One critical guideline is using proper PPE, including gloves, masks, and suits designed to prevent exposure to hazardous chemicals and drugs. In the event of contamination, decontamination procedures must be strictly followed to prevent the spread of harmful substances. Providing clear instructions on the correct usage and maintenance of PPE is a fundamental part of responder training.

Another guideline involves educating responders on the symptoms of chemical exposure and initial treatment protocols. Early recognition of symptoms such as respiratory distress, dizziness, or skin irritation can lead to prompt medical intervention, reducing health risks. For example, recognizing opioid intoxication symptoms like constricted pupils and depressed breathing guides responders in administering the appropriate dose of naloxone, potentially saving lives.

Furthermore, it's important to conduct regular drills and simulations that mimic real-life scenarios in illicit drug labs. These exercises enhance preparedness by allowing responders to practice identifying chemicals and executing protective measures under pressure. Continuous education on new chemical threats and evolving drug production methods ensures that responders remain capable of adapting to changing circumstances.

Response to Weapons of Mass Destruction (WMD) and Unknown Substances

The response to weapons of mass destruction (WMD) and unknown substances in hazardous materials incidents is an essential aspect of the work undertaken by HazMat technicians. WMDs include a wide array of destructive tools such as chemical, biological, and radiological agents that can drastically impact public health and infrastructure. Recognizing these threats requires technicians to develop keen vigilance and recognition skills.

Chemical agents like sarin or mustard gas can have immediate toxic effects on populations and are potentially devastating when used in concentrated attacks. For example, the Tokyo subway sarin attack demonstrated the effectiveness of chemical agents in civilian settings, causing numerous casualties and long-term health complications. Biological threats, such as anthrax, may not always be immediately apparent but have the potential to spread rapidly and cause significant harm if not promptly identified and contained (Gosden & Gardener, 2005). Radiological hazards, while less common, pose severe risks, particularly in the form of "dirty bombs" that disperse radioactive isotopes over large areas, necessitating prompt detection and thorough decontamination efforts.

A critical component of handling these threats is the need for heightened awareness and advanced pattern-recognition skills among first responders and HazMat technicians. This involves being able to quickly identify signs of a WMD attack, which might include multiple individuals displaying similar symptoms indicative of exposure to a toxic agent. The necessity of this vigilance stems from the varied means through which WMDs can be deployed, ranging from bombs to discrete dispersion methods involving food and water contamination (Biological and Chemical Terrorism: Strategic Plan for Preparedness and Response, 2019).

In addressing unknown substances, HazMat technicians must approach every situation with the precautionary principle in mind. This principle advises responders to assume the worst-case scenario until evidence indicates otherwise. For instance, the presence of an unmarked container or unusual substance at an incident site should be treated as potentially hazardous until confirmed safe. Such an approach mitigates risks, protecting both emergency personnel and the general public from unforeseen dangers.

Developing effective response strategies also involves coordinating closely with federal agencies and specialized teams. The vast expertise required to manage WMD incidents often exceeds the capabilities of local responders alone, making collaboration indispensable. Agencies like the Centers for Disease Control and Prevention (CDC) provide crucial support through their integrated disease surveillance systems and diagnostic laboratories, enhancing national preparedness against bioterrorism threats (Biological and Chemical Terrorism: Strategic Plan for Preparedness and Response, 2019).

Moreover, interagency cooperation ensures that responses are comprehensive and timely. Federal Emergency Management Agency (FEMA)'s operational plans outline collaboration frameworks that guide local responders in working alongside federal partners, ensuring resources are efficiently distributed and utilized during crises. Coordinating across different levels of government allows for more effective communication and mobilization of necessary technological and physical resources.

In tackling threats posed by WMDs and unknown substances, specific guidelines are vital in certain areas, such as radiation detection instruments and monitoring techniques. Technicians must be proficient in using Geiger counters and other devices to identify and assess radiological hazards efficiently. Training in these technologies helps ensure that responders can accurately measure environmental radioactivity and implement protective measures accordingly.

Similarly, decontamination procedures for radiation exposure must follow established protocols to prevent secondary contamination and safeguard public health. These procedures often include isolating affected individuals, removing contaminated clothing, and administering appropriate antidotes or treatment as needed.

Protective strategies for HazMat technicians are equally critical, encompassing the use of personal protective equipment (PPE) tailored to specific threats. Appropriate gear—ranging from full-body suits to respiratory masks—offers essential protection during hazardous operations, reducing the risk of exposure to dangerous substances.

The complexity of incidents involving WMDs and unknown substances underscores the importance of preparedness and adaptability among HazMat technicians. Training programs and simulation exercises play a fundamental role in honing the skills required to confront these challenges effectively. Scenarios that mimic real-world conditions help technicians practice not only technical skills but also evaluate decision-making processes under pressure.

Summary and Reflections

This chapter has provided a comprehensive exploration of technician-level responses to the unique challenges posed by hazardous materials incidents, especially those involving special hazards and terrorism-related threats. We delved into the intricacies of biological, radiological, and chemical threats, stressing the importance of understanding these agents' nature and behavior. Specialized training, sophisticated equipment, and interagency collaboration are key components that enable responders to effectively manage such complex situations. By acquiring knowledge about pathogen transmission, radiation protection, and chemical safety protocols, technicians can minimize risks and safeguard public health.

Moreover, as the threat landscape continues to evolve due to global terrorism and environmental changes, adaptability remains a critical attribute for HazMat technicians. Their role in coordinating with agencies and employing advanced technologies ensures efficient hazard management and effective response strategies. This chapter emphasized the necessity of continuous education and practice to prepare for unknown and emerging threats. By fostering an environment of learning and collaboration, technicians can maintain readiness and enhance their capacity to protect communities from the dangers associated with hazardous materials incidents.

20 Questions with Answers Based on Chapter 9: Special Hazards and Terrorism Response (Technician Level)

What is the main challenge when responding to HazMat incidents involving terrorism?

Answer: The main challenge is the unpredictability and complexity of terrorism-related hazards, which can involve biological, radiological, and chemical threats.

Why is specialized training essential for HazMat technicians when responding to terrorist threats?

Answer: Specialized training is essential because these threats require specific knowledge about different agents, their transmission, and containment methods to ensure responders' safety and effective management of the incident.

What are the three primary categories of threats covered in this chapter?

Answer: Biological, radiological, and chemical hazards.

How do biological agents, such as anthrax, pose a risk to responders?

Answer: Biological agents can spread through inhalation, ingestion, or skin contact, requiring responders to use proper protective equipment and follow containment protocols.

What are the three types of radiation discussed in the chapter?

Answer: Alpha, beta, and gamma radiation.

Why is it important for responders to understand different types of radiation?

Answer: Different radiation types require different protective measures, such as clothing for alpha particles and dense materials like lead for gamma rays.

What is the role of detection devices in HazMat responses to radiological threats?

Answer: Detection devices are used to assess environmental risks and determine the necessary personal protective equipment (PPE) levels for responders.

How do chemical threats like toxic industrial chemicals (TICs) pose challenges for responders?

Answer: Chemical threats like chlorine or ammonia require specific handling techniques and protective equipment to avoid exposure and contamination.

What is the significance of personal protective equipment (PPE) in responding to hazardous materials incidents?

Answer: PPE is crucial for protecting responders from exposure to harmful substances, such as chemical agents, radiation, or biological threats.

What role do CBRN units play in HazMat responses?

Answer: CBRN units provide specialized expertise and equipment to identify hazardous agents, conduct decontamination, and manage complex situations involving chemical, biological, radiological, and nuclear threats.

Why is interagency coordination important during HazMat incidents involving terrorism?

Answer: Interagency coordination enhances resource sharing, joint decision-making, and efficient management of complex situations, ensuring a comprehensive response.

How does global terrorism impact the nature of HazMat threats?

Answer: Terrorists may employ unconventional weapons or improvised explosive devices, introducing new and evolving threats that HazMat teams need to prepare for.

What are the main risks associated with illicit drug labs, particularly in the context of the opioid crisis?

Answer: Illicit drug labs can involve volatile chemicals, such as anhydrous ammonia and lithium, that pose risks of explosion, chemical exposure, and fire.

Why is fentanyl particularly dangerous for responders?
Answer: Fentanyl is highly potent, with even small amounts posing a severe risk of overdose and death through inhalation, ingestion, or skin contact.
What precautions should responders take when handling fentanyl?
Answer: Responders must wear appropriate PPE, including gloves, masks, and suits, to prevent exposure, and follow decontamination procedures to avoid contamination.
How does law enforcement support HazMat responses at illicit drug labs?
Answer: Law enforcement ensures scene security, supports evidence collection, and enforces legal protocols, enhancing safety and order at the site.
What is the "precautionary principle" when responding to unknown substances?
Answer: The precautionary principle advises responders to assume the worst-case scenario until the substance is identified and confirmed to be safe.
What specific equipment is required for radiation detection during HazMat incidents?
Answer: Equipment such as Geiger counters and dosimeters are used to detect and measure radiation levels, helping responders assess the risk and implement protective measures.
Why is collaboration with federal agencies critical in handling Weapons of Mass Destruction (WMD)?
Answer: Federal agencies provide advanced analytical capabilities, large-scale decontamination resources, and critical support, enhancing the local response to WMD incidents.
What is the importance of simulation drills and scenario-based training for HazMat technicians?

Answer: These drills allow technicians to practice identifying hazards, using equipment, and making decisions under pressure, ensuring preparedness for real-life incidents.

Chapter 10: Post-Incident Considerations and Responder Safety

(Technician Level)

Addressing post-incident considerations and responder safety is essential to effective hazardous materials (HazMat) management. The focus on this aspect ensures that the aftermath of a HazMat incident is managed with precision, securing both the immediate and long-term safety of responders and the community. Through meticulous post-incident procedures, actions are not only geared towards resolving the current situation but also toward preventing future incidents. This approach provides a structured framework for learning from each experience, enhancing preparedness, and adapting response strategies to evolving challenges in hazardous environments.

Within the chapter, readers will explore critical components designed to maintain responder safety and improve protocols continuously. It delves into systematic documentation practices that enable accurate incident reporting and provide a comprehensive account for legal compliance and procedural analysis. Further emphasis is placed on health monitoring to detect early signs of any adverse effects resulting from exposure, alongside robust training modules crafted to boost responder readiness and mental resilience. The evaluation process through after-action reports allows for reflection on the effectiveness of response strategies, recommending adjustments where necessary. In addition, the chapter discusses the importance of collaborative efforts within command structures to ensure efficient communication and coordination during incidents, thereby optimizing the effectiveness of the response while prioritizing the well-being of those on the frontline.

Introduction to Post-Incident Considerations

In the aftermath of any Hazardous Materials (HazMat) incident, the implementation of effective post-incident procedures is integral to ensuring the long-term safety of both responders and the community. These procedures serve as a backbone for not just immediate recovery but also future preparedness. The importance of these measures cannot be overstated, as they embody the collective learnings from past experiences and set the stage for improved response strategies.

One fundamental aspect of post-incident procedures is maintaining ongoing safety. This involves systematic documentation, thorough health monitoring, drawing lessons from the event, and continuous training—each playing a vital role in securing the well-being of individuals involved and those living within affected areas. Documentation serves as an initial step in post-incident procedures. It provides a detailed account of the incident, including chronological events, materials involved, and actions taken by responders. Such records are crucial not only for legal and regulatory purposes but also for evaluating the efficiency of response protocols. A well-documented incident offers valuable insights that contribute to refining safety procedures and adapting response strategies to mitigate potential risks in future operations.

Health monitoring stands out as another critical component. HazMat incidents often involve hazardous substances that pose short and long-term threats to responder health. Monitoring ensures that any adverse effects are detected early, allowing for timely medical intervention. Regular health screenings, aligned with occupational health standards, help identify chronic conditions like respiratory problems or other diseases that may arise from exposure to harmful agents. By implementing robust health surveillance programs, organizations can safeguard their workforce while enhancing overall response capabilities.

The process of extracting lessons learned from an incident is invaluable for continuous improvement. Through systematic evaluation, responders can identify what worked effectively and what did not. This analysis informs the development of After-Action Reports, which document insights and provide recommendations for improving future responses. These reports play a pivotal role in formalizing the learning process, ensuring that the same mistakes are not repeated. Learning from past incidents strengthens organizational resilience and fosters a culture of proactive risk management.

Training is the cornerstone of effective post-incident procedures. It ensures that responders are adequately prepared to handle future incidents with confidence and competence. Training programs should be dynamic, incorporating lessons learned from previous responses and adjusting to emerging challenges. Continuous training not only enhances technical skills but also prepares responders psychologically, equipping them to manage stress and maintain performance under pressure.

HazMat technicians and incident command structures play pivotal roles in executing these procedures seamlessly. Technicians are at the forefront, applying their expertise to manage hazardous situations efficiently. They ensure that safety protocols are followed meticulously and that all actions are documented accurately. Meanwhile, the incident command system coordinates efforts across different teams, providing a structured approach to managing complex situations. Effective communication and collaboration between these entities are essential for a coordinated response that minimizes risk and maximizes safety.

Equally important is addressing the mental, physical, and emotional well-being of responders. HazMat incidents can exert immense pressure on individuals, leading to stress, fatigue, and even trauma.

Supporting their well-being through psychological interventions, counseling, and peer support systems is crucial to maintaining a healthy workforce. By addressing these aspects, organizations ensure that responders are not only physically fit but also mentally resilient, capable of performing optimally in high-stress environments.

Documentation and Reporting Requirements

Accurate documentation plays a pivotal role in post-incident procedures, serving as a cornerstone for legal, regulatory, and safety measures within the realm of hazardous materials (HazMat) incident management. The requirement for precise documentation is not only a legal mandate but also a fundamental aspect ensuring accountability and transparency. The meticulous recording of data serves as evidence that all necessary precautions were taken during an incident, thus protecting the rights of both the responders and their organizations.

Firstly, precise documentation ensures compliance with legal requirements, creating an auditable trail that can withstand scrutiny. It provides proof that established protocols were followed, which is essential in case any future legal challenges arise. For instance, when incidents involve toxic chemicals, precise records are crucial to defend actions taken during emergency responses (<i>Working Relationships between OSHA and EPA | OSHA.gov | Occupational Safety and Health Administration</i>, 2025). Documentation safeguarding responder actions enhances transparency, reassuring stakeholders that every effort was made to minimize risks and uphold safety standards.

Comprehensive incident reports are integral to effective documentation. These reports must encapsulate detailed information—times of events, materials involved, and specific actions undertaken. The level of detail documented directly correlates with the report's usefulness. Ideally, these records should be completed promptly post-incident while memories are fresh, ensuring accuracy and reliability. Timely documentation of events allows for a clear understanding of what transpired, facilitating easier analysis and review.

In addition to basic details, comprehensive reports should include narrative sections that explain the context of the incident and the decisions made at each juncture. Such narratives bridge the gap between raw data and human judgment, offering insights into why certain actions were prioritized over others. This supports transparency and aids in training future responders by providing real-world scenarios that illuminate decision-making processes.

Additionally, communication with relevant authorities such as the Occupational Safety and Health Administration (OSHA), the Environmental Protection Agency (EPA), and other entities is mandatory. Each organization has specific regulatory requirements concerning incident reporting, and accurate documentation fulfills these obligations. By submitting precise reports, agencies like OSHA and EPA can maintain oversight of safety practices, ensuring that they align with regulatory standards and best practices (<i>Document, Document, Document: Dealing with an OSHA Citation before It Occurs</i>, n.d.).

Moreover, regulatory agencies depend on this documentation to evaluate compliance and investigate potential violations. Inadequate or inaccurate documentation could lead to hefty fines or legal penalties, thereby underscoring the necessity for thoroughness and attention to detail. Consistent and clear communication ensures that no steps are missed, fostering an environment of ongoing collaboration and improvement among responders and authorities alike.

After-Action Reports (AARs) form another vital component of accurate documentation. AARs serve the dual purpose of analyzing response performance and recommending improvements for future incidents. By reflecting collectively on what went well and identifying areas needing enhancement, responders can continually refine their strategies. This continuous learning process is essential in environments where conditions can vary dramatically from one incident to the next. The value of AARs lies in their structured approach to performance evaluation. They typically encompass several key elements: identifying what happened, what was supposed to happen, and gaps between these two understandings. Facilitating group discussions enables responders to voice perspectives, contributing to a more holistic understanding of the incident dynamics. It's this shared reflection that produces actionable insights, ultimately benefiting future preparedness and operational efficacy.

Medical Surveillance and Responder Health Monitoring

Health surveillance is a critical component in minimizing health risks associated with HazMat incidents, playing a vital role in ensuring responder safety. HazMat incidents inherently pose various health risks due to exposure to hazardous materials and environments. This exposure emphasizes the necessity for protective gear and stringent decontamination procedures. Proper protective equipment prevents direct contact with toxic substances, while thorough decontamination removes contaminants from responders' bodies and equipment, significantly reducing health risks.

Medical surveillance programs are essential in monitoring the health of responders regularly. These programs involve several key components designed to track and maintain responder health. Baseline health checks establish each responder's health status before exposure, providing crucial data for comparison after an incident. Post-incident physicals assess any immediate health implications following exposure to hazardous materials. Additionally, mental health evaluations are integral, addressing the psychological impact such incidents can have on responders. Mental well-being is often overlooked but is equally as important as physical health, given that stress and trauma can profoundly affect individuals involved in stressful or dangerous situations.

Early detection of chronic conditions resulting from HazMat exposure is paramount to maintaining responder health. Respiratory issues and cancer are among the chronic conditions that may arise from prolonged exposure to hazardous environments. Regular medical check-ups increase the chances of catching these conditions early, allowing for timely intervention and treatment. Continuous health monitoring provides a comprehensive view of responders' long-term health trajectory, aiding in identifying patterns or anomalies that could indicate emerging health issues.

Post-incident follow-up is a crucial part of health surveillance, involving regular health screenings to ensure compliance with occupational health standards. These screenings help in tracking any delayed health effects that may manifest over time, ensuring responders' continued fitness for duty. Moreover, adherence to established occupational health standards ensures that responders receive consistent, high-quality care aligned with regulatory requirements and industry best practices. Such standards guide the development of health protocols, ensuring they are robust and effective.

Guidelines for implementing medical surveillance programs, personal health management, and post-incident follow-up are necessary for structured and effective health monitoring. These guidelines should outline the frequency of health checks, specify procedures for handling identified health issues, and detail mental health support strategies. By following comprehensive guidelines, organizations can create a standardized approach to health monitoring, ensuring all responders benefit from systematic oversight.

The importance of health surveillance cannot be overstated. It ensures that responders remain healthy and capable of performing their duties, thereby safeguarding public health and safety. Comprehensive surveillance programs not only protect individual health but also contribute to overall operational readiness. When responders know their health is being monitored and cared for, it boosts morale and enhances their engagement and performance.

Furthermore, organizational commitment to health surveillance reflects a broader dedication to safety culture. Demonstrating a proactive approach to health risks encourages personnel to prioritize safety in all aspects of their work. This commitment to health and safety can also improve organizational reputation, establishing trust with both responders and the communities they serve.

Lessons Learned from Past Incidents

In the realm of hazardous materials (HazMat) response, each incident provides valuable opportunities for learning and enhancement. By meticulously analyzing past incidents, HazMat responders can identify common challenges, establish best practices, and refine future response efforts. This process of reflection and analysis is crucial for developing a more robust and efficient response to potential hazards.

One effective method for examining these incidents is through the study of case studies. Analyzing case studies allows responders to explore detailed accounts of previous incidents, giving them insight into the complexities and challenges encountered during HazMat operations. These case studies serve as a foundation for identifying recurring problems, such as the miscommunication between agencies or delays in response time, which could hinder an effective response. For instance, the CapWIN initiative highlighted the importance of interagency communication by showcasing its use during planned events like the Presidential Inauguration (<i>(PDF) Interagency Collaboration Challenges in Emergency Response</i>, n.d.). By understanding these barriers, responders can anticipate similar challenges in future scenarios and work towards overcoming them.

Evaluating response procedures and protocols is another critical aspect of enhancing the effectiveness of HazMat responses. By assessing existing strategies, responders can pinpoint areas that necessitate refinement or improvement. A systematic evaluation often reveals inefficiencies or outdated practices that require updating to align with current technological advancements and methodologies. For example, the phased implementation of the Wisconsin animal identification initiative emphasized step-by-step enhancements that controlled costs and minimized disruptions (<i>(PDF) Interagency Collaboration Challenges in Emergency Response</i>, n.d.). Such evaluations can guide responders in adopting phased improvements, ensuring that changes do not overwhelm the system but rather enhance its efficiency over time.

After-Action Reports (AARs) play an instrumental role in documenting insights gained from past incidents. These reports formalize the process of capturing lessons learned, allowing responders to systematically assess their performance and recommend actionable improvements for future training. AARs serve as comprehensive records that detail what worked well, what didn't, and why certain decisions were made. This documentation is invaluable not just for training purposes but also as a reference for policy changes and protocol adjustments. It ensures that all personnel involved are informed about the latest procedures and best practices, ultimately contributing to more effective responses.

Strengthening interagency collaboration is essential for improving overall response capabilities during complex incidents. Collaborative efforts among different agencies maximize resource utilization and ensure a coordinated approach to incident management. As seen in the BioSense initiative, expanding focus beyond regional crises to include naturally occurring disease outbreaks required enhanced coordination among various stakeholders (<i>(PDF) Interagency Collaboration Challenges in Emergency Response</i>, n.d.). This example underscores the benefit of seamless communication and joint exercises. By fostering relationships across agencies and conducting joint drills, teams can simulate real-life scenarios and practice working together effectively, thereby increasing their readiness for actual incidents.

To achieve these improvements, guidelines should be put in place for analyzing past incidents, evaluating current procedures, developing thorough after-action reports, and enhancing interagency collaboration. Each guideline aims to create a structured approach to incorporating lessons learned, ensuring that they translate into tangible advancements in HazMat response strategies.

Insights and Implications

In this chapter, we have delved into the essential post-incident procedures that are crucial for enhancing HazMat responder safety and operational efficiency. Emphasizing the importance of thorough documentation, this section highlighted how accurate records serve as a key component for legal compliance and strategic improvement. We explored the significance of maintaining ongoing health surveillance programs to safeguard responders from potential health risks associated with hazardous materials exposure. This proactive approach ensures early detection and intervention, ultimately promoting responder well-being and readiness.

Furthermore, we examined how extracting lessons learned from past incidents contributes to continuous improvement in response strategies. By systematically analyzing previous experiences, responders can identify effective practices and areas needing enhancement, leading to refined protocols. Training, as discussed, plays a vital role in preparing responders for future challenges, ensuring confidence and competence in high-pressure environments. The collaborative efforts between technicians and incident command structures emphasize seamless communication, underpinning coordinated responses that prioritize safety and effectiveness. As we conclude, it's clear that these post-incident actions not only strengthen individual and organizational resilience but also advance overall community safety.

20 questions with answers based on Chapter 10: Post-Incident Considerations and Responder Safety:

What is the primary purpose of post-incident procedures in HazMat management?

Answer: The primary purpose is to ensure long-term safety for responders and the community, improve preparedness, and adapt response strategies based on the lessons learned from previous incidents.

Why is documentation critical after a HazMat incident?

Answer: Documentation ensures legal compliance, accountability, and transparency, and provides valuable insights for improving future responses and safety protocols.

How does health monitoring contribute to responder safety after a HazMat incident?

Answer: Health monitoring helps detect early signs of adverse effects due to exposure to hazardous materials, enabling timely medical intervention and ensuring the responder's long-term health.

What role do after-action reports (AARs) play in post-incident procedures?

Answer: AARs analyze the response performance, identify what went well and areas for improvement, and provide recommendations to refine future response strategies.

Why is continuous training important for responder's post-incident?

Answer: Continuous training ensures that responders remain prepared, confident, and competent to handle future HazMat incidents, while also enhancing their psychological resilience.

What types of health issues can arise from prolonged exposure to hazardous materials?

Answer: Respiratory issues, cancers, and other chronic conditions may develop due to prolonged exposure to hazardous materials in HazMat incidents.

How does documentation help with legal and regulatory compliance after a HazMat incident?

Answer: Documentation creates an auditable trail that proves established protocols were followed, which is essential for legal defense and meeting regulatory requirements from entities like OSHA and EPA.

What should comprehensive incident reports include?

Answer: These reports should detail times of events, materials involved, actions taken, and provide a narrative explaining the context and decision-making processes during the incident.

How does medical surveillance benefit responders?

Answer: Medical surveillance tracks responders' health detects chronic conditions early, and ensures compliance with occupational health standards, ultimately maintaining responders' fitness for duty.

What is the importance of mental health evaluations after a HazMat incident?

Answer: Mental health evaluations are crucial for addressing the psychological impact of HazMat incidents, such as stress and trauma, which can affect responder performance and well-being.

How do after-action reports contribute to future HazMat response strategies?

Answer: AARs provide structured feedback on response performance, helping identify gaps or inefficiencies and guiding improvements in protocols, procedures, and training for future incidents.

What is the role of incident command structures in post-incident procedures?

Answer: Incident command structures coordinate the efforts of various teams, ensuring efficient communication, collaboration, and management of the incident and its aftermath.

Why is health surveillance important even after the incident is over?

Answer: Health surveillance continues post-incident to track delayed health effects, such as cancers or respiratory conditions, ensuring responders' long-term well-being.

How can analyzing past HazMat incidents improve response efforts?

Answer: Analyzing past incidents identifies common challenges, recurring issues, and successful practices, helping responders refine strategies and protocols for better future performance.

What is the value of training responders based on lessons learned from past incidents?

Answer: Training based on lessons learned ensures responders are better equipped to handle similar challenges, improving both their technical and psychological readiness.

What are the benefits of interagency collaboration during a HazMat incident?

Answer: Interagency collaboration maximizes resource utilization, ensures a coordinated response, and strengthens overall response capabilities during complex incidents.

What types of conditions are considered when establishing baseline health checks for responders?

Answer: Baseline health checks establish a responder's health status before exposure to hazardous materials, providing data to compare against post-incident evaluations for any health impacts.

How does regular health screening contribute to responder safety?

Answer: Regular health screenings help detect chronic conditions early, such as respiratory problems or cancers, ensuring responders receive timely medical care and remain fit for duty.

Why should post-incident debriefing include collaborative discussions?

Answer: Collaborative discussions during post-incident debriefing allow responders to share their perspectives, identify issues that were not apparent, and provide insights that lead to actionable improvements.

What is the role of personal health management in responder safety after a HazMat incident?

Answer: Personal health management includes taking proactive measures to monitor and maintain health post-incident, ensuring responders stay fit for duty and mitigating long-term health risks.

Book Summary: "Hazardous Materials Unveiled: The Foundations of Safety and Response"

Author: Eric J. Neal

Overview:

"Hazardous Materials Unveiled: The Foundations of Safety and Response" by Eric J. Neal is a comprehensive guide designed to equip responders, professionals, and those pursuing certification in hazardous materials (HazMat) safety with the knowledge needed to handle hazardous material incidents effectively and safely. This book is an invaluable resource for individuals at various levels of emergency management, fire services, and other sectors that respond to HazMat incidents.

Eric J. Neal uses a detailed, systematic approach to unravel the complexities of HazMat response and safety, with a particular focus on the practical application of protocols, documentation, responder health monitoring, and collaborative communication strategies. The content is designed not only to inform readers but also to serve as a study tool for individuals preparing for state certification exams and ongoing professional development. With a focus on real-world application, each chapter breaks down the critical concepts, best practices, and legal requirements, offering both theoretical foundations and hands-on strategies that are essential in dealing with HazMat incidents.

The book addresses everything from the basics of hazardous materials and their classification to the complexities of post-incident response, health monitoring, and inter-agency collaboration. It provides a step-by-step framework to prepare responders to safely manage and mitigate the risks associated with HazMat exposure.

Conclusion:

"Hazardous Materials Unveiled: The Foundations of Safety and Response" is a vital resource that equips readers with essential knowledge about hazardous material safety and incident management.

Eric J. Neal has crafted this book to not only inform readers but also to serve as a practical guide for professionals in the field, helping them better prepare for certification exams, job requirements, and real-world HazMat incidents. Through the clear explanation of protocols, safety measures, and the importance of continuous learning, this book stands as a foundational tool in the world of hazardous materials response.

Reference List

Argyropoulos, C. D., Obasi, I. C., Akinwande, D. V., & Ile, C. M. (2024, January 1). *The impact of interventions on health, safety and environment in the process industry*. Heliyon; Elsevier BV; Sciencedirect. https://www.sciencedirect.com/science/article/pii/S2405844023108127

EHS - NFPA Hazard Diamond | myUSF. (n.d.). Myusf.usfca.edu. https://myusf.usfca.edu/environmental-health-safety/nfpa-hazard-diamond

Environmental, Ml. (2017, November 21). *Hazardous Material Classification*. MLi Environmental. https://mlienvironmental.com/blog/classes-of-hazardous-materials/

Hazardous Waste Operations and Emergency Response (HAZWOPER) - Standards | Occupational Safety and Health Administration. (2025). Osha.gov. http://www.osha.gov/emergency-preparedness/hazardous-waste-operations/standards

Hazardous Waste Operations and Emergency Response (HAZWOPER) - Overview | Occupational Safety and Health Administration. (2025). Osha.gov. http://www.osha.gov/emergency-preparedness/hazardous-waste-operations

Nine Classes of Hazardous Materials (Yellow Visor Card) | FMCSA. (n.d.). Www.fmcsa.dot.gov. https://www.fmcsa.dot.gov/regulations/enforcement/nine-classes-hazardous-materials-yellow-visor-card

Nate. (2024, July). *Top Lessons Learned from Industrial Accident Disaster Cases*. JJ Legal. http://jjlegal.com/blog/lessons-from-industrial-accident-disaster-cases/

Smith, B. (2015). *GHS Hazard Classifications in Relation to HMIS® III & NFPA 704 Rating Systems*. Waxie.com. https://info.waxie.com/blog/ghs-hazard-classifications-in-relation-to-the-hmis-iii-nfpa-704-rating-systems

(2025). Labelmaster.com. https://www.labelmaster.com/hazmat-source/hazmat-labeling-placarding-marking?srsltid=AfmBOoqinw6iWWZY3aT38Tn0HoompA8DukyV2JDPq

DOT Hazard Classes. (n.d.). Www.iafc.org. https://www.iafc.org/topics-and-tools/hazmat/fusion-center/transportation-commodities/dot-hazard-classification-system

EHS - NFPA Hazard Diamond | myUSF. (n.d.). Myusf.usfca.edu. https://myusf.usfca.edu/environmental-health-safety/nfpa-hazard-diamond

FMCSA. (2013, December 29). *How to Comply with Federal Hazardous Materials Regulations.* Federal Motor Carrier Safety Administration. https://www.fmcsa.dot.gov/regulations/hazardous-materials/how-comply-federal-hazardous-materials-regulations

First Responder Technologies | Homeland Security. (2023). U.S. Department of Homeland Security. https://www.dhs.gov/archive/science-and-technology/first-responder-technologies

NFPA 704 Symbols - Institutional Risk & Safety | The University of Texas at Dallas. (n.d.). https://risk-safety.utdallas.edu/fire-prevention-and-protection-programs/nfpa-704-symbols/

Osha. (2024). *Safety Management - Hazard Prevention and Control | OSHA.gov | Occupational Safety and Health Administration.* Osha.gov. http://www.osha.gov/safety-management/hazard-prevention

US EPA. (2019, February 21). *Hazardous Waste Transportation | US EPA.* US EPA. https://www.epa.gov/hw/hazardous-waste-transportation

(2023). Fema.gov. https://emilms.fema.gov/is_0200c/groups/471.html

Savoia, E., Lin, L., & Viswanath, K. (2013, September). *Communications in Public Health Emergency Preparedness: A Systematic Review of the Literature.* Biosecurity and Bioterrorism: Biodefense Strategy, Practice, and Science. https://doi.org/10.1089/bsp.2013.0038

The Golden First Minutes — Initial Response to a Chemical Hazardous Materials Incident - CHEMM. (n.d.). Chemm.hhs.gov. https://chemm.hhs.gov/detailedinfo.htm

The First Responder: Technically Speaking. (2025). Aristatek.com. https://www.aristatek.com/Newsletter/05%2001%20January/The%20First%20Responder%20Technically%20Speaking.htm

Traffic Incident Management In Hazardous Materials Spills In Incident Clearance: 1.0 Introduction - FHWA Office of Operations. (2025). Dot.gov. https://ops.fhwa.dot.gov/publications/fhwahop08058/10.htm

Tulane University. (2023, January 18). *Developing a Disaster Communication Plan.* Publichealth.tulane.edu. https://publichealth.tulane.edu/blog/developing-disaster-communication-plan/

US EPA,OLEM. (2013, May 17). *Safety Zones | US EPA.* US EPA. https://www.epa.gov/emergency-response/safety-zones

Williams, J., Freeman, C. L., & Goldstein, S. (2021). *EMS Incident Command System.* PubMed; StatPearls Publishing. https://www.ncbi.nlm.nih.gov/books/NBK441863/

1926.65 - Hazardous waste operations and emergency response. | Occupational Safety and Health Administration. (2019). Osha.gov. http://www.osha.gov/laws-regs/regulations/standardnumber/1926/1926.65

CDC. (2023, May 5). *Donning and Doffing PPE: Proper Wearing, Removal, and Disposal.* Www.cdc.gov. https://www.cdc.gov/niosh/learning/safetyculturehc/module-3/8.html

California Code of Regulations, Title 8, Section 5192. Hazardous Waste Operations and Emergency Response. (2014). Ca.gov. https://www.dir.ca.gov/title8/5192.html

IN CASE OF HAZARDOUS SPILL: FIRST RESPONDER ROLE & INITIAL RESPONSE ACTIONS. (2025). Jensen Hughes.

https://www.jensenhughes.com/insights/in-case-of-emergency-first-responder-initial-response-actions

OSHA Technical Manual (OTM) - Section V: Chapter 3 | Occupational Safety and Health Administration. (2025). Osha.gov. http://www.osha.gov/otm/section-5-construction-operations/chapter-3

Personal Protective Equipment (PPE) - CHEMM. (2021). Hhs.gov. https://chemm.hhs.gov/ppe.htm

SeventhQueen. (2021, June 22). *A guide to the four levels of Hazardous Materials (HazMat) response.* HazmatNation. https://www.hazmatnation.com/a-guide-to-the-four-levels-of-hazardous-materials-hazmat-response/

Traffic Incident Management In Hazardous Materials Spills In Incident Clearance: 4.0 Hazard Materials Incident Clearance Compliance Requirements - FHWA Office of Operations. (2025). Dot.gov. https://ops.fhwa.dot.gov/publications/fhwahop08058/40.htm

7.4 Routes of Chemical Entry. (n.d.). Ehs.cornell.edu. https://ehs.cornell.edu/book/export/html/1381

7.4 Routes of Chemical Entry | Environment, Health and Safety. (n.d.). Ehs.cornell.edu. https://ehs.cornell.edu/research-safety/chemical-safety/laboratory-safety-manual/chapter-7-safe-chemical-use/74-routes

Abubakar, I. R., Maniruzzaman, K. M., Dano, U. L., AlShihri, F. S., AlShammari, M. S., Ahmed, S. M. S., Al-Gehlani, W. A. G., & Alrawaf, T. I. (2022). *Environmental Sustainability Impacts of Solid Waste Management Practices in the Global South*. International Journal of Environmental Research and Public Health.

Chemical Storage: Categories, Hazards And Compatibilities | Protocol. (n.d.). Www.jove.com. https://www.jove.com/v/10380/chemical-storage-categories-hazards-and-compatibilities

Hazard Communication - Guidance For Hazard Determination | Occupational Safety and Health Administration. (2023). Osha.gov. http://www.osha.gov/hazcom/ghd053107

Hazardous Waste Operations and Emergency Response (HAZWOPER) - Overview | Occupational Safety and Health Administration. (2025). Osha.gov. http://www.osha.gov/emergency-preparedness/hazardous-waste-operations

Mussorfiti, T., & Mussorfiti, T. (2022, June 9). *The Risk-Based Approach to Hazardous Material/WMD Incidents - Domestic Preparedness.* Domestic Preparedness. https://domprep.com/articles/the-risk-based-approach-to-hazardous-material-wmd-incidents

Team, T. S. (2024, December 10). *Simple But Needed.* Sbnsoftware.com. https://sbnsoftware.com/blog/what-are-the-environmental-impacts-of-improper-hazardous-materials-management/

2.2. Incidents on Land. (2022, December 6). Fema.gov. https://www.fema.gov/cbrn-tools/key-planning-factors-chemical-incident/kpf4/2/2-2

American Chemical Society. (1995). *Guide for Chemical Spill Response.* American Chemical Society. https://www.acs.org/about/governance/committees/chemical-safety/publications-resources/guide-for-chemical-spill-response.html

Document Display (PURL) | NSCEP | US EPA. (2025). Epa.gov. https://nepis.epa.gov/Exe/ZyPURL.cgi?Dockey=9101NNF0.TXT

Fletcher, B. (2023, November 10). *Technical Rescue: Hazmat Product Control for the Operations-Level Responder.* Firehouse.com; Firehouse. https://www.firehouse.com/rescue/article/53075652/how-firefighters-not-hazmat-technicians-control-hazardous-material-spills

Hazardous Waste Operations and Emergency Response (HAZWOPER) - Standards | Occupational Safety and Health Administration. (2025). Osha.gov. http://www.osha.gov/emergency-preparedness/hazardous-waste-operations/standards

Hazardous Waste - Decontamination | Occupational Safety and Health Administration. (n.d.). Www.osha.gov. http://www.osha.gov/hazardous-waste/decontamination

OSHA Field Safety and Health Management System (SHMS) Manual - Chapter 26 | Occupational Safety and Health Administration. (2025). Osha.gov. http://www.osha.gov/shms/chapter-26

29 CFR Part 1910 Subpart H—Hazardous Materials. (2015). Ecfr.gov. https://www.ecfr.gov/current/title-29/subtitle-B/chapter-XVII/part-1910/subpart-H

Document Display (PURL) | NSCEP | US EPA. (2025). Epa.gov. https://nepis.epa.gov/Exe/ZyPURL.cgi?Dockey=9100UJVC.TXT

Hazmat Implementing Response (Product Control Part 1) Flashcards. (2025). Quizlet. https://quizlet.com/547861077/hazmat-implementing-response-product-control-part-1-flash-cards/

OSHA Technical Manual (OTM) - Section VIII: Chapter 1 | Occupational Safety and Health Administration. (2025). Osha.gov. http://www.osha.gov/otm/section-8-ppe/chapter-1

SeventhQueen. (2021, June 22). *A guide to the four levels of Hazardous Materials (HazMat) response.* HazmatNation. https://www.hazmatnation.com/a-guide-to-the-four-levels-of-hazardous-materials-hazmat-response/

bradskramer. (2020, January 17). *What is the difference between hazmat awareness, operations, and technician levels for HAZWOPER?* Provenio. https://provenioconsulting.com/what-is-the-difference-between-hazmat-awareness-operations-and-technician-levels/

https://www.facebook.com/profile.php?id=1091021082. (2019). *HAZMAT Protection Levels - National Environmental Trainers®.* Natlenvtrainers.com. https://www.natlenvtrainers.com/blog/article/hazmat-protection-levels

(2024). Fema.gov. https://emilms.fema.gov/is_0100c/groups/25.html

5 keys to a great fireground incident action plan. (2017). FireRescue1. https://www.firerescue1.com/cod-company-officer-development/articles/5-keys-to-a-great-fireground-incident-action-plan-b05ZAXk2zCogHZw6/

Emergency Preparedness and Response: Getting Started | Occupational Safety and Health Administration. (2015). Osha.gov. http://www.osha.gov/emergency-preparedness/getting-started

Federal Register :: Request Access. (n.d.). Unblock.federalregister.gov. https://www.federalregister.gov/documents/2024/02/05/2023-28203/emergency-response-standard

Mussorfiti, T., & Mussorfiti, T. (2022, June 9). *The Risk-Based Approach to Hazardous Material/WMD Incidents - Domestic Preparedness.* Domestic Preparedness. https://domprep.com/articles/the-risk-based-approach-to-hazardous-material-wmd-incidents

NIMS Components - Guidance and Tools | FEMA.gov. (n.d.). Www.fema.gov. https://www.fema.gov/emergency-managers/nims/components

Role of Transportation Management Centers in Emergency Operations Guidebook - Section 3 Pre-Event Activities. (n.d.). Ops.fhwa.dot.gov. https://ops.fhwa.dot.gov/publications/fhwahop12050/sec3.htm

Team, T. S. (2024, December 7). *Simple But Needed.* Sbnsoftware.com. https://sbnsoftware.com/blog/how-do-you-conduct-a-hazardous-materials-risk-assessment/

<div class="hanging-indent"><i>Biological and Chemical Terrorism:Strategic Plan
for Preparedness and Response</i>. (2019). <a href="https://www.cdc.gov/mmwr/preview/mmwrhtml/rr4904a1.htm">https://www.cdc.gov/mmwr/preview/mmwrhtml/rr4904a1.htm</a></div>

Bland, S. A. (2013, November 22). *Chemical, Biological, Radiological and Nuclear (CBRN) Casualty Management Principles.* Conflict and

Catastrophe Medicine. https://doi.org/10.1007/978-1-4471-2927-1_46

- *FENTANYL: THE NEXT WAVE OF THE OPIOID CRISIS.* (n.d.). Www.govinfo.gov. https://www.govinfo.gov/content/pkg/CHRG-115hhrg25507/html/CHRG-115hhrg25507.htm

Gosden, C., & Gardener, D. (2005, August 11). *Weapons of mass destruction—threats and responses.* BMJ. https://doi.org/10.1136/bmj.331.7513.397

Kumar, V., Goel, R., Chawla, R., Silambarasan, M., & Sharma, R. (2010). *Chemical, biological, radiological, and nuclear decontamination: Recent trends and future perspective.* Journal of Pharmacy and Bioallied Sciences. https://doi.org/10.4103/0975-7406.68505

Thompson, J., Rehn, M., Lossius, H. M., & Lockey, D. (2014, September 24). *Risks to emergency medical responders at terrorist incidents: a narrative review of the medical literature.* Critical Care. https://doi.org/10.1186/s13054-014-0521-1

Tomassoni, A. J., Hawk, K. F., Jubanyik, K., Nogee, D. P., Durant, T., Lynch, K. L., Patel, R., Dinh, D., Ulrich, A., & D'Onofrio, G. (2017, February 3). *Multiple Fentanyl Overdoses — New Haven, Connecticut, June 23, 2016.* MMWR. Morbidity and Mortality Weekly Report. https://doi.org/10.15585/mm6604a4

1926.65 - Hazardous waste operations and emergency response. | Occupational Safety and Health Administration. (2019). Osha.gov. http://www.osha.gov/laws-regs/regulations/standardnumber/1926/1926.65

Document, Document, Document: Dealing with an OSHA Citation Before it Occurs. (n.d.). Hendershot Cowart P.C. https://www.hchlawyers.com/blog/2018/november/document-document-document-dealing-with-an-osha-/

Emergency Responder Preparedness Perceptions Concerning Hazmat Releases in the South Atlantic Region - ProQuest. (2020). Proquest.com. https://search.proquest.com/openview/

bd58cb1ea4a8f6cf2f978f16b4474ffa/
1?pq-origsite=gscholar&cbl=18750&diss=y

Hazardous Waste Operations and Emergency Response (HAZWOPER) - Standards | Occupational Safety and Health Administration. (2025). Osha.gov. http://www.osha.gov/emergency-preparedness/hazardous-waste-operations/standards

(PDF) Interagency Collaboration Challenges In Emergency Response. (n.d.). ResearchGate. https://www.researchgate.net/publication/271272916_Interagency_Collaboration_Challenges_In_Emergency_Res

Responder Safety and Health. (2024). ASPR TRACIE. https://asprtracie.hhs.gov/technical-resources/72/responder-safety-and-health

Working Relationships between OSHA and EPA | OSHA.gov | Occupational Safety and Health Administration. (2025). Osha.gov. http://www.osha.gov/laws-regs/mou/1991-02-13

Don't miss out!

Visit the website below and you can sign up to receive emails whenever Eric J. Neal publishes a new book. There's no charge and no obligation.

https://books2read.com/r/B-A-OJSQC-OQZAG

BOOKS 2 READ

Connecting independent readers to independent writers.

Did you love *Hazardous Materials Unveiled*? Then you should read *The Roadmap to Fire Behavior*[1] by Eric J. Neal!

[2]

In *The Roadmap to Fire Behavior*, Fire Chief Eric J. Neal offers an essential guide for firefighters, emergency responders, and fire science professionals. With decades of experience and expertise, Chief Neal delves into the science behind fire behavior, providing readers with a clear understanding of the principles that govern how fires ignite, grow, and spread.

This comprehensive resource examines critical topics such as the fire triangle and tetrahedron, heat transfer mechanisms, fire spread dynamics, and the impacts of building construction on fire behavior. It also explores the dangers of flashovers, backdrafts, and

1. https://books2read.com/u/b5KLkR

2. https://books2read.com/u/b5KLkR

ventilation-controlled fires, equipping firefighters with the knowledge to predict and adapt to rapidly changing conditions on the fireground.

Blending science with practical application, Chief Neal integrates real-world insights, case studies, and advanced technologies like fire modeling and thermal imaging to illustrate how understanding fire dynamics can save lives, protect property, and enhance safety.

Whether you are a new recruit, a seasoned firefighter, or an aspiring fire officer, this book serves as a vital tool for mastering the complexities of fire behavior and elevating your firefighting skills to the next level. *The Roadmap to Fire Behavior* is not just a book—it's a commitment to safety, preparation, and excellence in the fire service.

About the Author

Eric J. Neal is a dedicated public safety professional with a strong background in fire service leadership, emergency response, and community engagement. With years of experience serving and protecting his community, he has demonstrated a commitment to excellence, teamwork, and public service. Neal is actively involved in training, program development, and outreach initiatives that strengthen both emergency preparedness and community resilience. His leadership reflects a passion for service, integrity, and making a lasting impact in the communities he serves.

www.ingramcontent.com/pod-product-compliance
Lightning Source LLC
LaVergne TN
LVHW022041200926
848481LV00003B/742
9798230355519